SpringerBriefs in Space Development

Series Editor

Joseph N. Pelton Jr., Arlington, VA, USA

More information about this series at http://www.springer.com/series/10058

Erik Seedhouse

Spaceports Around the World, A Global Growth Industry

INTERNATIONAL®
SPACE UNIVERSITY
ISU

Springer

Erik Seedhouse
Commercial Space Operations
Embry-Riddle Aeronautical University
Daytona Beach, FL, USA

ISSN 2191-8171 ISSN 2191-818X (electronic)
SpringerBriefs in Space Development
ISBN 978-3-319-46845-7 ISBN 978-3-319-46846-4 (eBook)
DOI 10.1007/978-3-319-46846-4

Library of Congress Control Number: 2016953838

Printed on acid-free paper

This Springer imprint is published by Springer Nature
The registered company is Springer International Publishing AG
The registered company address is: Gewerbestrasse 11, 6330 Cham, Switzerland

This Springer book is published in collaboration with the International Space University. At its central campus in Strasbourg, France, and at various locations around the world, the ISU provides graduate-level training to the future leaders of the global space community. The university offers a two-month Space Studies Program, a five-week Southern Hemisphere Program, a one-year Executive MBA and a one-year Master's program related to space science, space engineering, systems engineering, space policy and law, business and management, and space and society.

These programs give international graduate students and young space professionals the opportunity to learn while solving complex problems in an intercultural environment. Since its founding in 1987, the International Space University has graduated more than 3,000 students from 100 countries, creating an international network of professionals and leaders. ISU faculty and lecturers from around the world have published hundreds of books and articles on space exploration, applications, science, and development.

Contents

Chapter 1
Spaceports: A Primer

Fig. 1.1 Virgin Galactic's SpaceShipOne. Image courtesy of Virgin Galactic

October 2004. The ink was barely dry on the $10 million Ansari X-Prize winning check before a potentially equally lucrative space race was announced; the competition between spaceports. Kick-starting the contest was Peter Mitchell, director of the New Mexico Office for Space Commercialization, who was present at Mojave Airport on the day that SpaceShipOne (SS1) made history. "Today doesn't belong

E. Seedhouse, *Spaceports Around the World, A Global Growth Industry*, SpringerBriefs in Space Development, DOI 10.1007/978-3-319-46846-4_1 to New Mexico, the day belongs to the gentlemen up here … that made this dream become a reality," Mitchell told reporters. "Tomorrow, however, we focus on bringing the spoils of this dream to the state of New Mexico". Not surprisingly, the remark rubbed some people the wrong way, including Dick Rutan, the brother of SS1 designer Burt Rutan, and a member of the Mojave airport district's board of directors. He promised to give Mitchell a run for his money and so the race was on (Fig. 1.1).

Remember, this was back in 2004, before the suborbital passenger business was even a business. But, with market studies predicting suborbital space tourism could generate more than billion dollars a year in revenues by the year 2021, the nascent industry's main players reckoned space passenger operations would require upgrades to attract the first wave of deep-pocketed thrill-seekers. After all, these passengers would need a place to train and a suitably high-end resort to stay. Not the sort of facilities you normally find at rocket-launch ranges, which generally have lots of wide-open space but little else. In short, there needed to be a viable tourist destination. And so the spaceport (Table 1.1) was born.

Table 1.1 FAA-licensed commercial spaceports in the United States

Spaceport name	Location	Operator	Services	Commercial license	Orbital	Sub-orbital
California Spaceport	Lompoc, CA	Spaceport Systems International	Payload processing	Issued 1996	Y	N
Mid-Atlantic Regional Spaceport	Wallops Island, VA	Virginia Commercial Space Flight Authority	Commercial, government, scientific, academic	Issued 1997	Y	Y
Kodiak Launch Complex	Kodiak, AK	Alaska Aerospace Corporation	Commercial, government	Issued 1998	Y	Y
Cape Canaveral Spaceport	Cape Canaveral, FL	Space Florida	Government, commercial, payload processing	Issued 1999	Y	Y
Mojave Air and Space Port	Mojave, CA	East Kern Airport District	Research and testing, commercial	Issued 2004	N	Y
Oklahoma Spaceport	Burns Flat, OK	Oklahoma Space Industry Development Authority	Commercial	Issued 2006	N	Y
Spaceport America	Las Cruces, NM	New Mexico Spaceport Authority	Commercial	Issued 2008	N	Y
Cecil Spaceport	Jacksonville, FL	Jacksonville Aviation Authority	Commercial	Issued 2010	Y	Y

The Federal Aviation Administration Office of Commercial Space Transportation (FAA AST) issues licenses to U.S. companies for commercial launches and they issue launch site licenses to spaceports

Fig. 1.2 Spaceport America. Image courtesy of Land Rover MENA

What is a Spaceport?

The truth is spaceports (Fig. 1.2) have been around for a while, but what is a spaceport? Well, it depends on the type of spacecraft and the type of people to some extent. Take the manned suborbital industry for example. We know this industry will depend upon two distinct kinds of space adventurer. First, there will be the wealthy few who can afford to buy a ($250,000 for suborbital) ticket and then there will be commercial scientist astronauts employed to conduct research and/or fly a payload. That's the first category. The second category of space adventurers are neither rich, nor do they have aspirations to be commercial astronauts, but they hope in turn to be able to afford the experience of a flight when prices come down. This second category will travel to spaceports to witness the experience and will want to feel involved vicariously in the flights. They will spend money on accommodation and food and drink and on souvenirs. So the most important thing is that they are able to get there. And, when they arrive, they will want to feel relaxed and welcome. An environment similar to a commercial airport, or a cruise ship terminal, will be needed.

Next are the training facilities (Fig. 1.3). Future spaceflight participants will need training facilities, which will be co-located at the spaceports. Centrifuges, hypobaric chambers, spatial disorientation trainers, classrooms, dunker training equipment; it all needs to be there. And, since some of this training will be stressful, it makes sense to co-locate medical facilities to check the health of the future spaceflight participants and certify them for space flight. This will be especially true in the early stages of the industry, because wealthy individuals, who can afford the flights, tend to be older and less healthy than average. There will also need to be emergency facilities in case of accidents.

Fig. 1.3 Training in ESA's Neutral Buoyancy Laboratory in Germany. Image courtesy of ESA

After a hard day's training, our astronauts-to-be and their friends will want to kick back and relax, so hotels will need to be built near, or attached to, the spaceports. Staying with the relaxation theme, it will make sense to co-locate entertainment facilities, so family and friends can occupy themselves during the training. Perhaps an IMAX type theater will be an attraction? Or a space theme park, with rides and space simulations perhaps? If these entertainment facilities are well designed, they could be a destination in themselves, even when there are no launches taking place. For example, the idea of a Space Camp/Academy is a great way to get kids involved and provide them with the opportunity to learn about the suborbital flight experience.

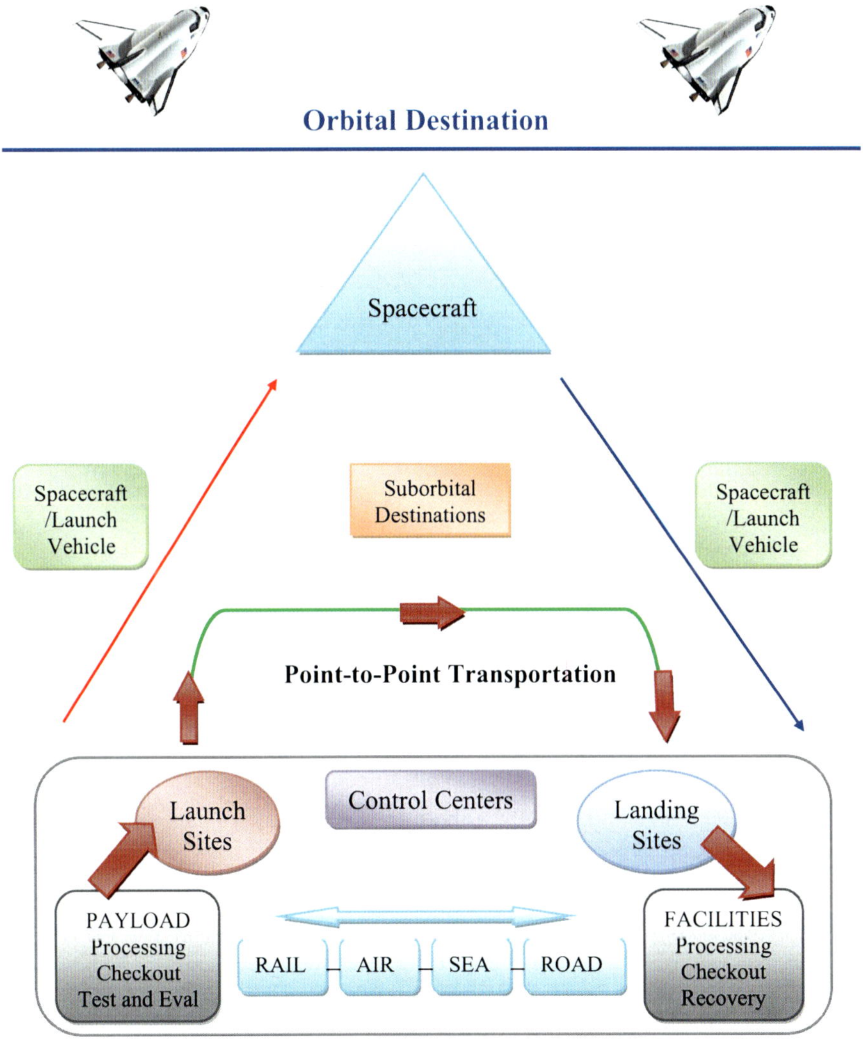

Fig. 1.4 Conceptual spaceport. Image by the author

For some, the term 'spaceport' (Fig. 1.2) may generate visions of sleek space-craft zipping between planets in a distant future but the fact is these facilities have been around for quite a while. Until very recently, spaceports were federally-owned facilities designed for launching big rockets into space. But, since the retirement of the Shuttle in 2011, NASA programs such as Commercial Crew and Cargo have created a pathway for private companies to muscle in on the lucrative space access business. And one of the elements of that business is a location from which to launch rockets: the spaceport. But different types of rockets require different types of spaceports: some launch vehicles such as the Atlas V require large scale steel structures while others such as the sounding rocket only need a small pad of concrete. And then there are the facilities to launch astronauts into space. In recent years much of the talk has been of spaceports servicing the impending manned suborbital industry as touted by Virgin Galactic, but the facilities servicing the launches undertaken by Paris Hilton and her celebrity friends will be a little different than those supporting professional astronauts. So, to make sense of just exactly what a spaceport is, it is useful to know what happens (Fig. 1.5) at such a facility and also what common elements comprise it.

What Happens at a Spaceport?

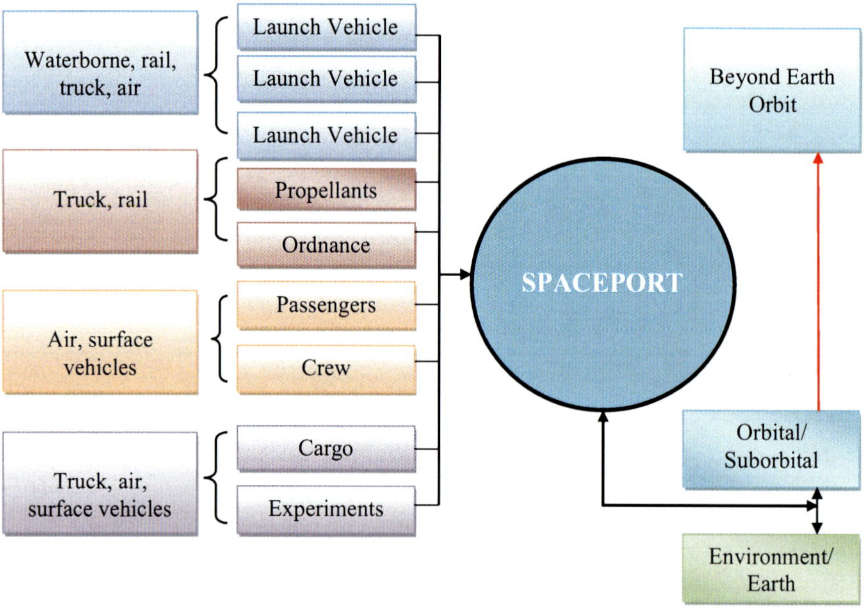

Fig. 1.5 What happens at a spaceport. A spaceport is a transportation hub where all the elements for a spaceflight come together. First, components arrive by air, rail, water or by road. Next, rocket elements are integrated to create the launch vehicle. Propellants are then transferred to the launch vehicle, the crew boards and the vehicle launches. Adapted from a diagram by Wayne Finger

Figure 1.5 shows the sequence of events that occur at a spaceport. We'll discuss these events in more detail later in this brief, but for now we just need to know that this sequence of events is a function of the five common components—*spaceport, control center and airspace, payload processing facilities, launch vehicles* and *spacecraft*—of a spaceport. These are illustrated in Table 1.2 using Florida Spaceport as an example.

Table 1.2 Florida spaceport facilities

System component	Definition	Florida assets
Spaceport 	A public gateway to space that typically provides launch and re-entry sites. In the U.S., launch facilities that serve commercial, non-governmental customers must be licensed by the Federal Aviation Administration (FAA).	• **Cape Canaveral Spaceport**: commercial facilities at Kennedy Space Center (KSC) and the 45th Wing at Cape Canaveral Air Force Station (CCAFS) • **Cecil Spaceport**: a newly licensed facility in western Jacksonville.
Control Centers and Airspaces	Centers that coordinate the details for space flight operations. Airspace in space transportation is primarily concerned with *ranges*, a flight path area used for launching rockets and vehicles designed to reach high altitudes.	• **Launch Control Center** (LCC) at KSC. • **Morrell Operations Center (MOC)** manages the 15-million square mile Eastern Range. • **Dedicated Launch Vehicle Control Centers** for the Atlas V, Delta IV, and Falcon 9.
Launch Vehicles and Spacecraft 	A *launch vehicle* is a rocket used to launch a spacecraft or satellite into high altitude/orbit. They are classified as reusable (RLVs) or expendable (ELVs). *Spacecraft* are manned or unmanned vehicles designed to operate in space to accomplish a mission.	• The **Atlas V**, **Delta IV**, and the **Falcon 9** that will launch from CCAFS. • Development of the **Space Launch System (SLS)** at KSC. • **Suborbital-ready facilities** at Cecil Spaceport.
Payload Processing Facilities 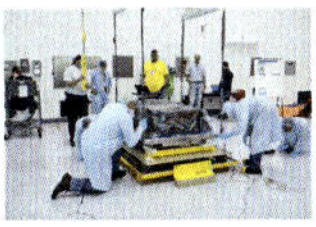	Facilities that prepare payloads (the cargo necessary to complete a mission or flight's purpose) for launch, and processing following the flight.	• **12 major facilities at KSC and CCAFSs** with the capability to process a variety of payload types and sizes. • **Astrotech** in Titusville
Intermodal Facilities 	Transportation modes that enable the movement of people and goods to spaceports, including roadways, airports, seaports, and rail lines.	• **Strategic Intermodal System (SIS),** a system of key roadway, rail, airport, seaports, and spaceport infrastructure identified by the Florida Department of Transportation.

Adapted from Florida Spaceport Plan 2013

Now that we have an understanding of what happens and what comprises a generic spaceport it is helpful to know a little about the types of launches and vehicles. We'll being with suborbital versus orbital. Suborbital launches (Fig. 1.6) barely reach space, which is defined as 100 km altitude, and they do not complete an orbit of the Earth. In the past, these types of launches were conducted by governments to test missiles, but with the arrival of Virgin Galactic and the development of suborbital reusable launch vehicles (sRLVs), these launches will likely increase over the next 10–15 years. In fact it is possible that by the mid-2020s weekly suborbital manned launches could be taking place.

Fig. 1.6 SpaceShipTwo. Image courtesy of Virgin Galactic

Suborbital vs. Orbital

From a spaceport perspective, suborbital launches don't require as much infrastructure as orbital launches simply because trips to the edge of space are simpler operations than trips that orbit the Earth. That said, there are many common elements and systems of the spaceport infrastructure that apply to suborbital and orbital, and each system (Table 1.2) and sub-system has a role to play. For example, on a broad level, Control Centers coordinate spaceflight operations, but this can only be achieved through the use of the sub-systems of range control, launch vehicle control, and telemetry and tracking, which must work together to ensure airspace is safe for launch and that vehicles can be monitored during flight. Another key element is coordination of the National Airspace System (NAS), through which spacecraft must fly (Fig. 1.7). Managing a spaceflight through the NAS is not an easy task when you consider the dynamic flight path of a rocket that might be traveling at Mach 3 or 4 vertically and that this flight path may intersect with horizontal flight corridors populated with commercial traffic flying along at 'only' 800 km h.

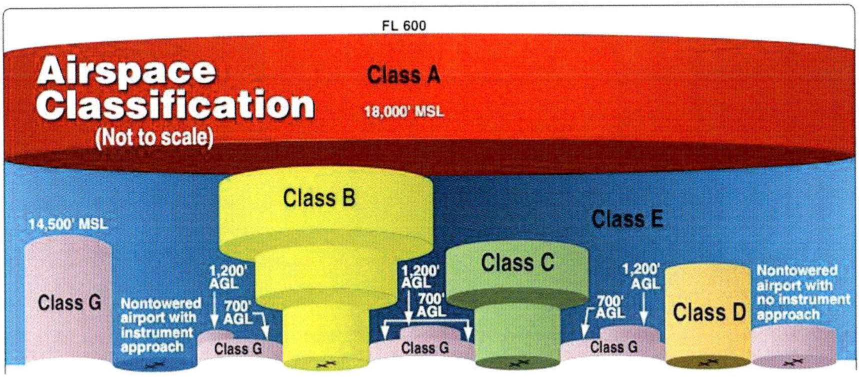

Fig. 1.7 Airspace classification. Image courtesy of FAA

Another key element of the spaceport are the launch vehicles, which come in all shapes and sizes. The variety and type of launch vehicles that can be supported by a spaceport will to a large extent dictate the customer base. So, for example, a spaceport that can only support expendable launch vehicles (ELV's) that launch only payloads will attract a different customer base than a spaceport that can support ELV's and reusable launch vehicles (RLV's) that launch payloads *and* astronauts. Equally, the types of launch vehicles supported by a spaceport will drive the fidelity of the payload processing facilities: in spaceport parlance, the payload is the cargo, which may be equipment, a satellite, a person, or a combination of these. Payload processing (Fig. 1.8), as we shall see in Chap. 4, varies considerably, depending on the features of the payload: human payloads require very different processes than a piece of equipment for example.

Fig. 1.8 Payload processing. Image courtesy of NASA

And then there is the business of getting to and from a spaceport: the intermodal connections required by a spaceport supporting unmanned vehicles will be a different transportation infrastructure than that required by one supporting manned operations for example. We'll discuss this later in the brief, the purpose and structure of which is outlined here.

Purpose of this Brief

The purpose of this brief is to provide a general overall understanding of what a spaceport comprises and what new technologies will be needed in the short-term to develop these facilities. As suborbital operators such as Blue Origin and Virgin Galactic, and orbital operators such as SpaceX and Sierra Nevada Corporation, ramp up towards manned revenue flights, spaceports are fast becoming the center-of-attention in the commercial spaceflight arena. Several spaceports have already been built, work is underway building new facilities, and many more are in the development phase. Around the world, from Curaçao to Kona, promises of future economic prosperity have led governments to provide financing and incentives to enhance existing spaceports and develop new ones. But the business needs of the nascent commercial spaceflight industry are significant and some existing spaceports fall short of being able to offer the complete package that satisfies the myriad user requirements. This brief explains what those business needs are, how spaceports as a private space enterprise can grow, what the promised economic benefits from this emerging industry may be, and how spaceports will be critical for developing the commercial spaceflight industry.

This brief also describes the three overarching goals that govern the direction of the current crop of spaceports, namely:

1. Delivering efficient and effective services to all customers, providing affordable, flexible, and fast-turnaround capabilities and facilities, and providing astronauts and spaceport visitors with a high quality spaceport experience.
2. Driving local job creation and injecting the economy with greater demand for goods, services and a skilled workforce that spans the gamut of trades and professions related to commercial spaceflight.
3. Inspiration. Spaceports can contribute to inspiring guests and the next generation of astronauts by boosting interest in STEM-related education.

Finally, this brief describes how spaceports can help broaden the scope of the commercial spaceflight industry by developing infrastructure to accommodate all manner of space launch activities, including horizontal and vertical launch and return concepts, near-space high altitude balloons such as WorldView, air-launched satellite systems such as LauncherOne, government-sponsored spaceflight initiatives, and military requirements.

Structure of this Brief

The first chapter of this book highlighted the key elements of a generic spaceport to provide the reader with a nuts-and-bolts account of what a spaceport is and an overview of the goals that govern the direction of a spaceport. Chapter 2 provides an insight into the history of spaceports while Chap. 3 delves into the safety and liability issues that govern spaceport operations.

In Chap. 4, the challenges of integrating launch vehicles into the National Airspace System (NAS) are described, along with a discussion of the technologies and strategies being developed to achieve this. Chapter 5 provides a brief overview of some of the launch vehicles being serviced by spaceports. This chapter also provides an insight into the emerging crop of commercial suborbital launch vehicles such as SpaceShipTwo (SS2) and New Shepard.

Chapter 6 examines the scope of payload processing facilities required at a spaceport. This is an important topic since spaceports must cater for all types of cargo, ranging from plasma physics to passengers and from biological payloads to biophysics experiments. Chapter 7 addresses facilities designed to support missions conducted by spaceflight participants, payload specialists and scientists. While the length of training for passengers will vary, many of the types of training will be same. For example, suborbital and orbital spaceflight participants will be required to undergo centrifuge training, high altitude indoctrination and parabolic flight training.

Chapter 8 addresses the challenges of point-to-point transportation (PTP) and why this mode of transportation is unlikely to be realized in the near future due to the lack of market demand and the technology challenges that are a long way from being realized. Chapter 9 concludes this book by featuring select spaceports around the world and looking over the horizon at spaceports currently being developed.

Chapter 2
Spaceports: A Definition and Brief History

Fig. 2.1 Spaceport America may not be the busiest spaceport on the planet, but it is definitely the most visually striking. Credit: Spaceport America/Spaceport America Conceptual Images by URS/Foster and Partners. Image courtesy of FAA

E. Seedhouse, *Spaceports Around the World, A Global Growth Industry*,
SpringerBriefs in Space Development, DOI 10.1007/978-3-319-46846-4_2

If you build it, they will come.
Field of Dreams, 1989

They had taken Moses Callahan's ship and turned it into paper.

A man lived on his ship. He breathed her air, ate and drank from her stores. Her bulkheads solid around him kept the uncaring vacuum outside where it belonged and her driving engines bent the very curvature of space to take him wherever he wanted to go.

But then he had to land. . . .

Suddenly all that breathing and eating became a life-support replenishment invoice. Those protecting bulkheads hid structural support members that had to be inspected and recertified by a licensed and commensurately expensive naval surveyor. Engines became fuel costs and a ten-thousand-hour service charge. Then there were berth fees, entry fees, value-added tax on cargo transactions, customs "courtesy" fees, outright bribes to the longshoremen's union—and Moses Callahan wound up sitting in the deepest corner of the Hybreasil inport bar complex, wondering whether to have another beer or have his good uniform cleaned and pressed before heading outport to try to unearth a cargo Celtic Crescent or Western Galactic might have overlooked.

From *The Shattered Stars* by Richard McEnroe, 1983

Spaceports in Science Fiction

Spaceports have been a mainstay of science fiction for decades (Fig. 2.1). Think of the spaceports in *Elysium* or *The Fifth Element*, or Mos Eisley, perhaps the most famous spaceport to grace the silver screen. Mos Eisley, as Star Wars fans will know, is the low-grade concrete structure that appears in the first film of the franchise. Home to the Mos Eisley Cantina, the Star Wars version of a spaceport includes a docking bay (Docking Bay 94) which houses the Millenium Falcon, and the aforementioned cantina which is the haunt of all manner of characters, including a band of alien musicians—the Figrin D'an and the Modal Nodes. But in the real world, what is a spaceport? Well, one way to answer that is to consider the types of people and service providers who make up the space industry and to review the features you might expect to see when you visit one of these facilities. We'll begin with the impending manned commercial spaceflight industry and consider what facilities will be needed by the Blue Origins and Virgin Galactics of this world. We already know this industry will depend upon the two distinct kinds of space adventurer mentioned in the previous chapter. These future spaceflight participants will need training facilities, which will be co-located at the spaceports. Centrifuges, hypobaric chambers, spatial disorientation trainers, classrooms, dunker training equipment; it all needs to be there (Table 2.1).

Basic Spaceport Facilities

Table 2.1 Basic spaceport features

Class	Feature description
Local Infrastructure	Runway
	Railhead
	Road access
	Hotels, restaurants and shops
	Qualified local workforce
	Proximity to university
Site Facilities	Pads for sounding rockets
	Pads for small, medium and large sRLVs
	Horizontal takeoff/landing capability
	Fuel handling/solid
	Fuel handling/liquid
	Fuel handling/hybrid
	Chemical analysis facilities
	Ordnance facilities
	Vehicle integration/checkout
	Payload processing-hazmats
	Processing—dynamic balance
	Spacecraft storage facilities
	Engineering/mission management offices
	Range radars, cameras
	Telemetry data retrieval
	Payload processing-vibration
	Engine test stands
	Materials testing facilities
	Hazmat training
	On-site research labs
	Broadband access
	Emergency response teams
	Downrange payload retrieval.
Space training	Medical facilities
	Training facilities
	Simulators
	Space academy
	Family facilities/residential
	Family facilities/entertainment
Financial/admin	Financial incentives/trade zones
	International facilities/customs
	Security for military users
	High tech company incubators
	Simplified admin (safety, environment)

So spaceports (see Appendix A for a full list of facilities) will be a centralized travel hub for spaceflight participants, but passengers aren't the only cargo being launched into space from these facilities. Satellites, payloads and science experiments are also launched from these spaceports. To get a better idea of just how versatile some of these spaceports might be it is instructive to examine today's macro space transportation system as depicted in Fig. 2.2.

As you can see, spaceports will support a range of functions and the spectrum of those functions will determine whether the spaceport is sparse and utilitarian or sprawling and cutting edge. But no matter how many facilities are co-located at a particular spaceport, one of the major drivers for success will be location. Ideally a spaceport will need to be at the center of a large network of transportation. Some of the current crop of US spaceports fit this requirement (Spaceport Florida for example) whereas others (Spaceport America) are far removed from a nexus of major roads or transport links of any kind.

Spaceport History and Development in the United States

In a 2008 AIAA paper [1], *Public-Private Spaceport Development*, G. Wayne Finger, P David L. Keller, and Brian S. Gulliver divided the history of US spaceports into distinct phases, the first of which was the *Pre-Spaceport Phase* that had its origin in the 1950s, when missile test ranges were needed to develop missile programs. These ranges tended to be located in uninhabited areas, but as missile capability grew, coastal locations became more popular since over-flight problems were eliminated. The *Pre-Spaceport Phase* transitioned to the *Federal Spaceport Phase* shortly after the creation of NASA in 1958. With NASA came a need for civilian launch capabilities and so began the gradual evolution of missile ranges at Cape Canaveral (Fig. 2.3) to orbital spaceports, which usually shared military facilities.

The *Federal Spaceport Phase* became the *Federal—State Mixed Spaceport Phase* in the 1980s, a phase characterized by military and commercial launches taking place from Government-built launch pads. This phase was also notable for the increased development of commercial launch vehicles that were launched from commercial launch sites, the development of Shuttle Launch Complex 8 (SLC-8) at Vandenberg Air Force Base (AFB) (Fig. 2.4) and the transition of SLC-46 by Spaceport Florida (sidebar).

As the 1980s gave way to the 1990s, the *Federal—State Mixed Spaceport Phase* transitioned to the *CONUS vs. OCONUS Phase*. It was during this period that US space launch policy changes resulted in the industry becoming much less competitive than it had in previous decades: between 1963 and 1980 the US government launched all the western world's commercial satellites for the simple reason that nobody else was in the business of launching payloads into space. But this changed in 1980 when

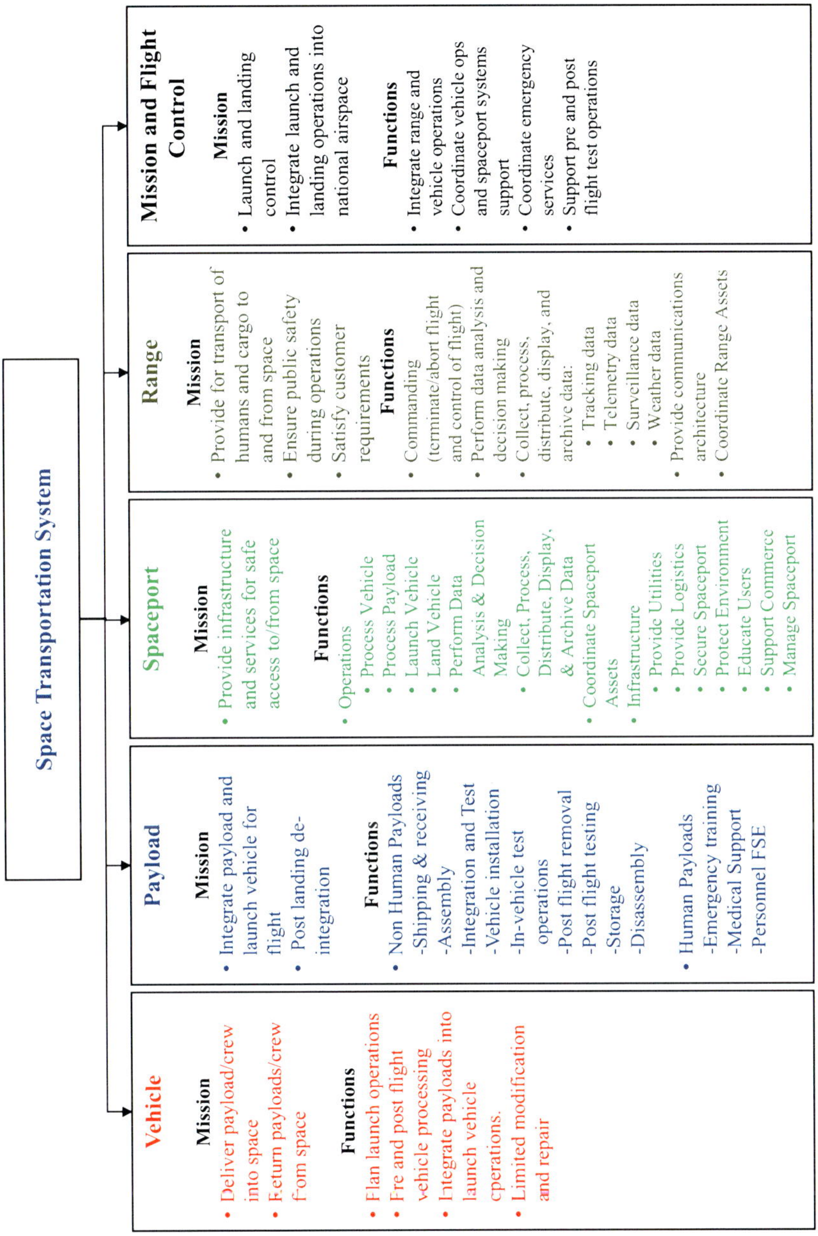

Fig. 2.2 Today's macro space transportation system. Image by the author

Fig. 2.3 Cape Canaveral. Image courtesy of NASA

the European Space Agency (ESA) came along with the Ariane (Fig. 2.5) expendable launch vehicle (ELV). ESA stated it aimed to launch 50 % of the commercial satellites, and the agency got off to a great start thanks to the 1984 US decision to shut down all ELV production and fly all commercial and government satellites on board the Space Shuttle. This plan may have worked had it not been for the Challenger accident in 1986, which forced the US government to reverse its 'no ELV' policy. But by this time ESA had had a six year head start in the ELV business and it wasn't until 1991, with the arrival of the commercial Atlas (Fig. 2.6) and Delta vehicles that a true commercial space industry came on line. By that time Ariane had captured a large chunk of the worldwide launch contracts, a situation that was made worse for the US by the emergence of Chinese and Russian ELVs.

In the same year that US ELV production was suspended, the Commercial Space Launch Act (CSLA) came into being (it was amended in 1988 and codified in 1994) and it has been this instrument that has provided guidance for facilitating US commercial space activities ever since. The CSLA also established the Department of Transportation (DoT) as the agency with oversight for the conduct of commercial launch operations. In this capacity, the DoT calls for the Secretary of Transportation to promote and facilitate commercial space activity and to encourage private sector involvement to expand and modernize space launch infrastructure. But back to the *CONUS vs. OCONUS Phase*. With the increased competition from ESA, China and Russia, new US vehicles sought launch sites outside of the continental United States (OCONUS), which included the development of a launch site in the Australian desert. By the early 2000s this phase gave way to the *Mixed and Customized Phase* which was notable for the appearance of suborbital adventure space tourism and the development of Virgin Galactic's SpaceShipOne.

Fig. 2.4 Vandenberg Air Force Base. Image courtesy of NASA

Fig. 2.5 ESA's Ariane launch vehicle. Image courtesy of ESA

Fig. 2.6 Atlas launch vehicle. Image courtesy of NASA

Spaceport Development

Since SpaceShipOne, Virgin Galactic has gone on to develop SpaceShipTwo and other companies such as Blue Origin are also in the space tourism game with the development of their launcher, New Shepard. The development of these launchers has, inevitably, had an impact on spaceport development (Table 2.1).

Table 2.2 Spaceport development

			Phase				
Spaceport	**Pre-Spaceport Role**	**Date of First Use as Space-port**	1950's - Pre-Spaceport	1960's - Federal	1980's - Federal & State Mixed	1990's - CONUS vs OCONUS	2000's - Mixed & Customized
Spaceport America (NM)	State-owned desert	2006					
Mojave Air and Spaceport (CA)	Mojave Airport	2004					
California Commercial Spaceport (CA)	Vandenberg AFB	1999					
Kodiak Launch Complex (AL)	Kodiak National Wildlife Refuge	1998					
Spaceport Florida (FL)	Cape Canaveral AFS	1997					
Poker Flatt Research Range (AL)	University of Alaska	1968					
Kennedy Space Center (FL)	Privately-owned land	1968					
Reagan Test Site (Kwajalein)	Naval Station Kwajalein	1959					
Vandenberg AFB (FL)	US Army Camp Cook	1957					
Cape Canaveral AFS	Banana River NAS	1950					
Wallops Flight Facility (VA)	NACA	1945					

Adapted from *Public-Private Spaceport Development*, by G. Wayne Finger, P David L. Keller, and Brian S. Gulliver (1)

As you can see in the above table, spaceports have been developed by US Federal agencies (NASA at Kennedy) with specific launch programs and spaceports have also been developed by entrepreneurial space companies (Virgin Galactic at Spaceport America). Along the way, the requirements that affect spaceport development (sidebar) have also changed. In the early phases Federal launch vehicles dominated the market and most of these launches ferried satellites into orbit, although a number of NASA launches also launched humans. The location of the spaceports for Federal Government launchers was driven by the following factors:

1. Proximity to the equator, since this increases payload weight to orbit.
2. Proximity to coast, since this reduces danger to life and property.
3. Proximity to military base, since this enhances security.

Spaceport Florida

Florida doesn't just have one spaceport, but a growing network that initially operated under the auspices of Florida Space Authority (FSA) that was created as a state government space agency in 1989. Mandated to expand and diversify Florida's space industry, the FSA, which technically is not an agency, was then consolidated with enactment of the Space Florida Act in 2006 and Space Florida was created. Four years later Space Florida was issued Real Property Licenses for Launch Complexes 36 and 46 at Cape Canaveral, a move that was followed in 2012 by the State of Florida requesting 150 acres of NASA land to be developed as a commercial spaceport. Known as 'Shiloh', this location will eventually become Shiloh Commercial Spaceport, thereby joining Cecil Spaceport in Jacksonville, which can support suborbital launches, and Kennedy Space Center, which signed over its management to Space Florida in 2013.

In the later phases, the factors driving spaceport location [2] changed because the suborbital space tourism market was a very different one from the federally-fueled ELV market. For one thing, the vehicles being developed by operators such as Blue Origin are designed to be reusable, which means it is not desirable to make a touchdown in the ocean because such a mission profile would decrease recovery options. Also, for operators such as Virgin Galactic, there is no compelling reason to co-locate with a military base because of the added burden of security and regulations, neither of which contributes to the profit margins. And, since these launches will be suborbital, locating the spaceport near the equator becomes less important because the bigger driver is a site that offers training facilities and hotel accommodation for guests and friends.

Spaceport History and Development in Russia

Spaceports go back a long way in Russia. All the way back to 1946. This was the year that a launch site was selected for testing missiles in Kapustin Yar, a village in the Astrakhan Oblast region on the lower Volga. The site (Fig. 2.7) became operational a year later, in October 1947, with the launch of a R0-1 ballistic missile. This launch was followed by many more ballistic missile launches between 1948 and 1956, under the leadership of S. P. Korolev. As the years went by the spaceport infrastructure was upgraded to include structures for vertical launches and accommodation for personnel to live onsite.

Fig. 2.7 Kapustin Yar spaceport. Image courtesy of ESA

Kapustin Yar

The Kapustin Yar location truly became a spaceport on 16th March 1962, with the launch of Kosmos-1. Then, on 14th October 1969, the spaceport assumed the status of international spaceport with the launch of Interkosmos-1. For many years following the Interkosmos-1 launch, Kapustin Yar functioned as a launch facility for small missiles and satellites. Then, in 1988, when the demand for small satellites fell sharply, Kapustin Yar was put on ice, although the complexes were maintained in good enough condition to launch the German spacecraft ABRIXAS and MegSat in 1996. Today the Kapustin Yar spaceport is operated by the Russian Ministry of Defense.

Baikonur

At about the same time Kapustin Yar was taking shape, the Soviets were busy developing another spaceport. Located about 2130 km (via the M32) to the east of Kapustin Yar, Baikonur (Figs. 2.8 and 2.9) is perhaps the most familiar spaceport to Westerners since this is where NASA astronauts are launched to the International Space Station (ISS). Baikonur had its origins in the 1950s when there was a requirement to support the development of the Soviet's first intercontinental ballistic missile, the R-7. After much deliberation, a site in Kazakhstan near the town of Kzyl-Orda was selected, and work went ahead to construct the necessary telemetry stations, launch control center, assembly test facility and the launch complex itself. Baikonur quickly became the most famous spaceport on the planet since it was the

Fig. 2.8 Buran at the Baikonur launch facility. Image courtesy of Taylor Empire Airways

Fig. 2.9 One of the busiest spaceports on the planet: Baikonur. Image courtesy of NASA

location from which the first artificial satellite and first human were launched. As the years passed the spaceport grew dramatically, with the addition of myriad launch and maintenance facilities to support light (Tsyklon), intermediate (Soyuz and Zenit), heavy (Proton—Fig. 2.10) and super heavy launchers.

Fig. 2.10 Proton launch vehicle. Image courtesy of NASA

Today, the Baikonur Spaceport (which was known as the Baikonur Launch Site until being renamed in 1990) is operated under the authority of the Russian Aviation and Space Agency, although the Russian Ministry of Defense controls one Proton launch complex.

Fig. 2.11 Plesetsk. Image courtesy of NASA

Plesetsk

Due to the rapid development of Soviet space activities and ambitions in the 1950s, the Soviets established a third spaceport in the northern region of the USSR. Construction of what was to become a launch site initially, began in Plesetsk (Fig. 2.11) in 1957. During the 1960s and 1970s the complex was used to prepare and launch Soyuz, Kosmos and Molniya vehicles. Today, Plesetsk, which was granted spaceport status in 1994, is the only missile launch site in Europe, although the only missiles launched are for defense and scientific purposes.

Vostochny

Russia's newest spaceport is the Vostochny Cosmodrome which, in 2016, is under construction on the 51st parallel north. Due for completion in 2018, Vostochny is being built to reduce Russia's dependency on the Baikonur Cosmodrome which is located in Kazakhstan. Located near the city of Tsiolkovsky, the spaceport's 51° N latitude location means rockets launched from the site will be capable of carrying almost the same payload weight as those vehicles being launched from Baikonur's

46° N location. But, more importantly, the Vostochny Spaceport will save the Russians money because of the yearly rent Russia pays to spaceports located in other countries (Russia pays Kazakhstan $115 million annual rent to use the facilities at the Baikonur Spaceport for example). In 2016 only about 25 % of Russian launches will actually take place on Russian territory, but with the new spaceport that figure is projected to rise to 90 % by 2030, with 45 % of Russian launches slated to be moved to Vostochny by 2020.

Spaceport History and Development in China

Jiuquan Satellite Launch Center

China's spaceport history stretches back to 1958 when the first of the country's spaceports was founded near Jiuquan, about 1600 km from Beijing. The Jiuquan Satellite Launch Center (JSLC) and spaceport, like all Chinese launch facilities, is remote and closed to foreigners. Modelled on Soviet counterparts, JSLC's launch complexes have been the epicenter of many of China's ventures into space, including the first Chinese manned mission, Shenzhou 5 in October 2003, and the country's first satellite (Dong Fang Hong 1) in 1970.

Taiyuan Satellite Launch Center

Founded 8 years after the JSLC, the Taiyuan Satellite Launch Center (TSLC) is a site used mainly for launching earth resource satellites, scientific satellites and weather satellites on Long March vehicles. Known also as Base 25, the spaceport also happens to be the site from which China's intercontinental ballistic missiles are tested.

Xichang Satellite Launch Center

More commonly known as the Xichang Space Center, the Xichang Satellite Launch Center (XSLC) spaceport was declared operational in 1984 and is used to launch geostationary and weather satellites. The spaceport gained international notoriety in 1996 when a Long March 3B rocket went off course just seconds after launch, crashing into a village, killing six and injuring 56. The spaceport was featured again in international news 11 years later when China launched its infamous SC-19 anti-satellite (ASAT) missile from the site: the SC-19 kinetic kill vehicle (KKV) impacted China's FY-1C weather satellite at an altitude of 865 km creating a serious space debris problem that will continue to present a hazard to orbiting satellites for decades. Given the XSLC's high latitude and proximity to inhabited areas, it is likely many of the geostationary satellite launches will eventually be transferred to the Wenchang Satellite Launch Centre (WSLC), although the XSLC will remain open for military launches.

Wenchang Satellite Launch Center

The WSLC is China's fourth spaceport. Formally a suborbital launch center, the WSLC, thanks to its low latitude location (19° N) will eventually be capable of launching the heavy lift Long March 5 booster, which means this spaceport will play a key role in China's future manned space program, space station and lunar program. One of the advantages of the WSLC is its proximity to a sea port for transport of the Long March's 5-m core boosters, which are too wide to be transported by rail.

Spaceports Around the World

Guyana Space Center

Outside the US, Russia and China, perhaps the most prominent spaceport is the European Spaceport (Fig. 2.12) located at Kourou in French Guyana which has been operational since 1968. Initially selected to become a French launch center, the spaceport became a shared facility with ESA in 1975. Since then ESA has paid two thirds of the budget and has also covered the cost of upgrades made to the Ariane launchers.

Fig. 2.12 Kourou spaceport. Image courtesy of ESA

Thanks to its location very close to the equator (latitude 5°10′ N) and with vast swathes of uninhabited territory stretching in all directions, Kourou is ideal for launching rockets to low inclination orbits, something it has done very successfully for decades. Although Kourou is best known for launching Ariane rockets (Fig. 2.13) it also supports launches of the Russian Soyuz-2 vehicle for which ESA built the ELS (l'Ensemble de Lancement Soyouz) pad. In the ESA-Russia venture, ESA is increasing its rocket family with Soyuz vehicles which it uses to launch commercial payloads. Russia meanwhile has access to Kourou, which it uses to launch its own payloads; this is a cost-saving measure for the Russians because the same Soyuz rocket that can launch only 1.7 tonnes into geostationary transfer orbit (GTO) from Baikonur, can launch 2.8 tonnes from Kourou.

Fig. 2.13 Ariane rocket. Image courtesy of ESA

Fig. 2.14 India's lunar orbiter Chandrayaan-1. Image courtesy of ISRO

Satish Dhawan Space Center

India's spaceport is the Satish Dhawan Space Center (SDSC), a former satellite launch center that became operational in 1971. Today the complex comprises two launch pads which support the launch of unmanned vehicles such as India's lunar orbiter Chandrayaan-1 (Fig. 2.14), launched in 2008, and the country's Mars orbiter, Mangalyaan, launched in 2013. With an eye on future manned launches, India is constructing a third launch pad.

The SDSC got its start with the launch of a small sounding rocket in 1971. Since then the site has been significantly upgraded to support orbital launches and vehicle integration. Today, the SDSC sports two orbital launch pads together with launch facilities for sounding rockets, telemetry and tracking facilities, a vehicle assembly complex, liquid propellant storage, and a payload clean room. Key to India's manned spaceflight aspirations are the Second Launch Pad that caters to future vehicles, and the Third Launch Pad, which is being constructed specifically for manned missions.

Andøya Space Center

Norway's spaceport is Europe's most northerly rocket launch site. It also happens to be a very busy facility, with more than 1200 sounding rockets having been launched since 1962. One of these sounding rockets gained particular notoriety in 1995, when the Russians mistook the launch for a nuclear missile launch from an American submarine. Fortunately a counter-strike was avoided. Ninety percent of the facility is owned by the Royal Norwegian Ministry of Trade and Industry and the remainder by Kongsberg Defence and Aerospace. In 2016 the facility is preparing to launch an orbital launcher capable of delivering a 10 kg payload in to polar orbit.

References

1. Wayne Finger, G., Keller, D.L., Gulliver, B.S.: Public-private spaceport development. In: SpaceOps 2008 Conference (Hosted and organized by ESA and EUMETSAT in association with AIAA), Reynolds, Smith & Hills, Inc., Merritt Island, FL, pp. 225–7201
2. Finger, G.W., Kercsmar, J.C., Gulliver, B.S.: Trends in spaceport site selection and operational requirements. In: International Space Development Conference (2006)

Chapter 3
The Regulatory Environment

Fig. 3.1 Navy cruiser Lake Erie launches a SM-3® IB missile. The SM-3® missile is an interceptor used by the U.S. Navy to destroy ballistic missile threats. The interceptor uses a kill vehicle to intercept (read: collide) with targets in space. Since the impact is the equivalent of a 5-tonne truck traveling at 1000 kmh there is no need for explosives. Credit: U.S. Navy

E. Seedhouse, *Spaceports Around the World, A Global Growth Industry*,
SpringerBriefs in Space Development, DOI 10.1007/978-3-319-46846-4_3

International Treaties

The legal environment in which spaceports operate and will operate is very much a work in progress because there are so many national, regional and supranational legal mechanisms that must be considered [1–4]. Not surprisingly, given the country's busy launch schedule, it is the United States that has the most advanced regulatory framework that regulates and legislates what happens at a spaceport. In the U.S., spaceport policy comes under the jurisdiction of the Federal Aviation Administration's Office of Commercial Space Transportation (FAA—AST), which developed the current legal spaceport environment by first examining civil aviation policies and the current crop of United Nations (UN) international treaties (Table 3.1/Appendix B) that concern space access (Fig. 3.1).

Table 3.1 United Nations (UN) treaties that govern space-related activity

➢ Outer Space Treaty.
➢ 1967 Treaty on Principles Governing the Activities of States in the Exploration and Use of Outer Space, Including the Moon and Other Celestial Bodies
➢ The 1968 Agreement on the Rescue of Astronauts, the Return of Astronauts and the Return of Objects Launched into Outer Space
➢ The 1972 Convention on International Liability for Damage Caused by Space Objects
➢ The 1975 Convention on Registration of Objects Launched Into Outer Space
➢ The 1979 Agreement Governing the Activities of States on the Moon and Other Celestial Bodies

While the term 'spaceport' is not mentioned in any treaty, there are some basic principles that apply to the development and operation of a spaceport, these being articles that pertain to astronaut safety, liability, ground safety, responsibility, and registration. In addition to the UN treaties, there are also national policies that have been applied to the business of developing spaceports as outlined below:

United States: The U.S. operates under the National Space Policy that came into effect in 2006 [5]. This policy includes the following two principles that apply specifically to spaceports:

(i) Ground segments of space capabilities are deemed vital to national interests.
(ii) The U.S. is committed to facilitating an entrepreneurial space sector.

More guidance can be found in the U.S. federal regulations Title 14, part 420 (Appendix C), which explains how a potential spaceport operator may apply for a launch license (sidebar).

This excerpt is provided to give readers a snapshot of the spaceport regulatory environment. In addition to a launch license, spaceports must also obtain a re-entry license.

Launch law[1] in the U.S. is codified in 14 Code of Federal Regulations (CFR) 401.5 as follows: *Launch* means to place or try to place a launch vehicle or reentry vehicle and any payload from Earth in a suborbital trajectory, in Earth orbit in outer space, or otherwise in outer space, and includes preparing a launch vehicle for flight at a launch site in the United States. Launch includes the flight of a launch vehicle and includes pre- and post-flight ground operations as follows:

(1) *Beginning of launch*

(i) Under a license, launch begins with the arrival of a launch vehicle or payload at a U.S. launch site.
(ii) Under a permit, launch begins when any pre-flight ground operation at a U.S. launch site meets all of the following criteria:

(A) Is closely proximate in time to flight,
(B) Entails critical steps preparatory to initiating flight,
(C) Is unique to space launch, and
(D) Is inherently so hazardous as to warrant the FAA's regulatory oversight.

(2) *End of launch*

(i) For launch of an orbital expendable launch vehicle (ELV), launch ends after the licensee's last exercise of control over its launch vehicle.
(ii) For launch of an orbital reusable launch vehicle (RLV) with a payload, launch ends after deployment of the payload. For any other orbital RLV, launch ends upon completion of the first sustained, steady-state orbit of an RLV at its intended location.
(iii) For a suborbital ELV or RLV launch, launch ends after reaching apogee if the flight includes a reentry, or otherwise after vehicle landing or impact on Earth, and after activities necessary to return the vehicle to a safe condition on the ground.

Spaceport Policy Country by Country

The content and intent of the aforementioned treaties has been applied to varying degrees by the countries that currently host spaceport facilities. What follows is a brief overview of the extent of regulation in countries that host spaceports.

Russia. All of Russia's space activities operate under the Russian Federal Space Program in accordance with the Space Act, but this policy is unclear on the subject of licensing and the development of new spaceports.

---

[1] U.S. law applies to governmental and nongovernmental space activities through, inter alia: λ Commercial Space Launch Act λ Land Remote Sensing Policy Act λ Communications Act of 1934 λ National Aeronautics and Space Act.

China. China has no official spaceport policy, although it conducts commercial launches at its three launch sites.

Australia. Spaceport regulation in Australia is covered by the Space Activities Act of 1998, which identifies policy goals and describes safety requirements and considerations. Australia's space policy also clearly defines the launch of a vehicle and facility and launch license regulations, although there is little mention of intermodality or coordination with the global marketplace.

Brazil. The Brazil Space Agency (AEB) outlines policy goals for the agency, which include encouraging the private sector, stimulating entrepreneurship, and international cooperation. The AEB has sound legal basis and regulates launch activities at launch sites by means of commercial launch licenses rather than granting site licenses.

Canada. The Canadian Aviation Regulations briefly mention the need for ministerial authorization when launching a rocket, but there is no mention of aligning entrepreneurial goals with spaceport facilities and operations. In Canada rocket launch license regulations are found in aviation regulations rather than domestic space law.

Indonesia. Indonesia relies on a Space System rather than a space act to regulate its space activities. This space system includes services, ground stations, infrastructure and spaceports.

Japan. There is very little regulation that pertains to licensing launch activities at spaceports in Japan, which relies on national space law to conduct it launch activities.

Kazakhstan. The regulations that pertain to cosmodromes such as Baikonur are very clear on the issues of safety, compliance with international obligations and global competition, but there is very little specific information that speaks to spaceports.

Russia. The Law on Space Activity, Section IV, Article 18, is the reference for the regulation of spaceport infrastructure but there is little information on the jurisdiction of activities conducted at a spaceport.

Ukraine. The Ukraine regulates and licenses sites through its domestic space law, which have a sound basis in law and are detailed enough to be implementable.

Europe. In Europe, the European Aviation Safety Agency (EASA) has proposed to certify suborbital vehicles as aircraft and to regulate spaceports as aerodromes [6]. EASA rules provide standard guidelines for management, spaceport operation, certification, and oversight. The rules do not have a completely sound legal basis since they have not been adopted by the European Commission but the aviation regulations pertaining to suborbital activity are detailed enough to be clear and effective.

Safety

To gain approval for a launch site location, an applicant shall demonstrate that for each launch point proposed for the launch site, at least one type of expendable or reusable launch vehicle can be flown from the launch point safely.[2]

In the U.S. spaceport safety is regulated through the CSLA by Congress, which in turn directs the FAA—AST to ensure public health and safety is protected [7–9]. So, to obtain a spaceport license, the applicant must undergo a thorough safety assessment that includes various quantitative risk analyses such as the one presented in Table 3.2.

Table 3.2 FAA-AST AC 431.35-2A risk matrix

Severity / Frequency	**Catastrophic I**	**Critical II**	**Marginal III**	**Negligible IV**
Frequent (A)	**1**	**3**	**7**	**13**
Probable (B)	**2**	**5**	**9**	**16**
Occasional (C)	**4**	**6**	**11**	**18**
Remote (D)	**8**	**10**	**14**	**19**
Improbable (E)	**12**	**15**	**17**	**20**

Level	**Index**	**Hazard Risk Acceptability Criteria**
High	1 - 6	Corrective/controlling actions must be taken to reduce the hazard severity below 'II' or reduce the likelihood of occurrence below 'C'
Medium	7 - 10	If not controlled, the risk must be accepted by Program Management and FAA
Low	11 - 20	Project Management decides on actions, if any

Until recently when it came to defining and managing (Fig. 3.2) acceptable risk for passengers, the only standards were those used by NASA for the Space Shuttle, which had a 1-in-65 risk of loss of mission and loss of crew [10]. Since such a high risk is unacceptable for the commercial spaceflight industry, the FAA-AST has provided Advisory Circular (AC) 437.55-1 that supplements the CFR requirements [11]. But while AC 437.55-1 defines acceptable risk, the FAA does not lay out any level of acceptable risk for *passengers*: commercial spaceflight is very much a 'fly at your own risk' endeavor. On the subject of *crew* safety, the FAA's requirements are a little more specific, stating that the crew must be able to control a RLV and be capable of acting in a contingency event.

[2] Title 14 → Chapter III → Subchapter C → Part 420 → Subpart B. §420.19 Launch site location review—general.

Spaceport Risk Management

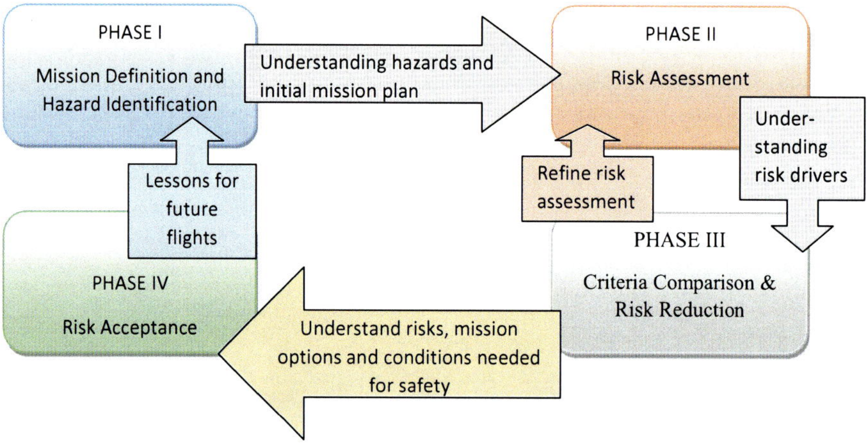

Fig. 3.2 Spaceport risk management process. Image by the author

In the U.S., the risk threshold for granting a spaceport license the probability of risk 10^{-6} is applied for Expected Casualty for third parties on the ground, although this threshold varies depending on whether vehicles are ELVs or RLVs.

Role of the Commercial Space Launch Act

The Commercial Space Launch Act (CSLA I 51 U.S.C. §§ 50901–50923), which was amended in 1989 and 1998, directs the FAA to promote and regulate the commercial space industry. This means the FAA (FAA-AST) is responsible for licensing commercial launch and reentry sites and ensuring commercial space transportation is safe, in addition to the following tasks:

- Safely open access to space and encourage private sector development
- Simplify and expedite issuance and transfer of launch and reentry licenses
- Promote safety
- Strengthen and expand space transportation infrastructure
- Provide licensing Requirements for:

  (i) Launch/reentry in U.S.
  (ii) Launch/reentry by U.S. citizen outside U.S.
  (iii) For launch/reentry by U.S. citizen outside U.S. and outside territory of foreign country, unless foreign country's government has an agreement with U.S. on jurisdiction over the launch or operation.
  (iv) For launch/reentry by U.S. citizen in foreign country if U.S. has jurisdiction by agreement with government of foreign country

- License applications subject to policy, safety and environmental impact reviews.
- Compliances with Registration Convention
- Orbital debris mitigation
- Flight crew qualifications, training, safety and waiver of claims against U.S.

International Traffic on Arms Regulations

Fig. 3.3 Yes, according to the provisions of ITAR and the USML, that suborbital vehicle you plan to take a trip on is classified as a weapon! Credit: U.S. Navy

The International Traffic in Arms Regulations, or ITAR, has been a ball and chain around the neck of U.S. commercial spaceflight for years [12]. ITAR was originally developed to regulate military products and services, but the regulations now cover many products that were initially developed for the military but have recently become commercially available. At the core of ITAR is the United States Munitions List, or USML, which lists weapons (Fig. 3.3, Table 3.3), military vehicles, flight control products and satellites, launch vehicles and ground control equipment. This category means that spacecraft such as Virgin Galactic's SpaceShipTwo (SS2) is classified as a munition, which means Virgin Galactic must exercise caution because any violation could result in criminal liability and/or imprisonment. It goes without saying that ITAR/USML has rubbed the commercial spaceflight industry the wrong way partly because of the broad category definition of Category XXI which includes the following catch-all: *any other product, software, service or technical data with substantial military capability that was designed, developed, configured, adapted or modified for a military purpose.*

Table 3.3 United States munitions list

I—Firearms	XII—Fire Control, Range Finder, Optical and Guidance and Control Equipment
II—Artillery Projectors	XIII—Auxiliary Military Equipment
III—Ammunition	XIV—Toxicological Agents and Equipment and Radiological Equipment
IV—Launch Vehicles	XV—Spacecraft Systems and Associated Equipment
V—Explosives, Propellants, Incendiary Agents and Their Constituents	XVI—Nuclear Weapons Design and Related Equipment
VI—Vessels of War and Special Naval Equipment	XVII—Classified Articles, Technical Data and Defense Services Not Otherwise Enumerated
VII—Tanks and Military Vehicles	XVIII—Reserved
VIII—Aircraft and Associated Equipment	XIX—Reserved
IX—Military Training Equipment	XX—Submersible Vessels, Oceanographic and Associated Equipment
X—Protective Personnel Equipment	XXI—Miscellaneous Articles
XI—Military Electronics	

Unfortunately the ITAR/USML issue extends beyond material, which Virgin Galactic has discovered to its cost. In 2012 Virgin Galactic was featured in the commercial space blogs with the news that the company was not allowed to fly Chinese nationals. Why? ITAR.

> Virgin Galactic adheres to both the spirit and the letter of US export controls and has for now chosen not to accept deposits from countries subject to US export and other regulatory restrictions.
>
> Virgin Galactic statement, January 31, 2014

This case was related to Part 126.1 of ITAR which prohibits the export of technologies under its control to selected nations, one of which is China. This had nothing to do with Virgin Galactic flying SS2 from China, something that would most definitely be prohibited under ITAR. The reason for not being able to sell tickets to the Chinese was that such an act would fall foul of the export regulations since the ticket would be classified as a 'related item' under the list of prohibited items on the USML. There was some progress in 2013 when Virgin Galactic's flight operations were removed from the control of ITAR. This meant that the company could fly non-US citizens without having to jump through the hoops necessary to get an export licence (under the old rules Sir Richard Branson wouldn't have been allowed to fly on his own spacecraft because he's British!). This appeared to be good news, but the exemption Virgin Galactic received stated that as long as the company's hardware was built in the United States, the government would still have authority over anything the company sent abroad.

Lobbying by the commercial spaceflight industry has had some effect. In December 2012 Congress struck out the legalese that placed satellites on the USML, although the prohibitions on export remained. But in its new draft Congress added 'man-rated suborbital, orbital, lunar and interplanetary spacecraft' to Category XV of the USML. It was a case of two steps forward one step back, with the result that the lobbying continued unabated. In April 2015 a commercial space advisory group submitted a proposal to remove manned commercial spacecraft from the USML at a meeting of the FAA's Commercial Space Transportation Advisory Committee. After much deliberation among members, the committee agreed to recommend that manned commercial suborbital vehicles should be removed from ITAR's jurisdiction. Whether Congress approves remains to be seen.

References

1. Hobe, S.: Legal aspects of space tourism. Neb. L. Rev. **86**(2) 441–458 (2007)
2. Masson-Zwaan, T., Freeland, S.: Between heaven and earth: the legal challenges of human space travel. Acta Astonaut. **66**, 1597–1607 (2010)
3. Mineiro, M.C.: Law and regulation governing U.S. commercial spaceports: licensing, liability, and legal challenges. J. Air L. & Com. **73**, 759 (2008)
4. Nield, G., et al.: Certification versus licensing for human space flight in commercial space transportation. Proceedings of the 63rd International Astronautical Congress, Naples, Italy (October 2012)
5. Galloway, E.: Law and security in outer space: the role of congress in space law and policy. J. Space Law **22**(1&2), 35–50 (1983)
6. European Aviation Safety Agency: Certification Specifications for Large Aeroplanes, CS-25, Book 2, Acceptable Means of Compliance, 27 December 2007
7. Mineiro, M.: Assessing the risks: tort liability and risk management in the event of a commercial human space flight vehicle accident. J. Air L. & Com. **74**, 371 (2009)
8. Quinn, A., et al.: New suborbital safety technical committee for the IAASS. Proceedings of the 5th IAASS Conference 'A Safer Space for a Safer World', Versailles, France, 17–19 October 2011 (ESA SP-699, January 2012)
9. Quinn, A.: Acceptable levels of safety for the commercial space flight industry. IAC-12-D6.1.2 Presented at 63rd IAC Naples, Italy, October 2012
10. Howard, D.: The elephant in the room: informed consent from the spaceport operator's perspective. Air & Space Law. **25**(2) (2012)
11. Federal Aviation Administration, Code of Federal Regulations, Title 14, Part 25.1309(b)
12. Mineiro, M.C: Space Technology Export Controls and International Cooperation in Outer Space. Space Regulations Library, Chapter 1, pp. 3–6 (2011)

Chapter 4
Control Centers and Airspace

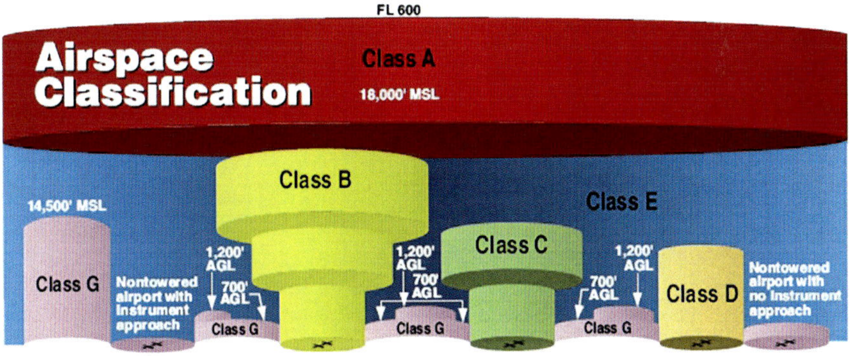

Fig. 4.1 Airspace classification. Image courtesy of Federal Aviation Administration

This chapter discusses how the U.S. controls airspace during rocket launches (Fig. 4.1). While other countries have procedures in place to control airspace during launch events, these are not as sophisticated as those administered by the Federal Aviation Administration (FAA), which has developed the most advanced means of integrating launch vehicles into its national airspace system.

Role of the FAA

Today, the U.S. space program utilizes civil, military and commercial sectors, the latest of these being the commercial sector, which had its start with the passage of the Commercial Space Launch Act (CSLA) in 1984. Regulatory oversight was then

E. Seedhouse, *Spaceports Around the World, A Global Growth Industry*,
SpringerBriefs in Space Development, DOI 10.1007/978-3-319-46846-4_4

passed to the FAA's Office of Commercial Space Transportation (FAA-AST), which is now tasked with protecting the public, property and national security interests of the U.S. during commercial launches. To achieve its mission the AST requires operators to obtain a license before launching a commercial vehicle and also to obtain a re-entry license (missions conducted by the government are exempt from this administrative requirement). As of this writing, the FAA has issued more than 250 launch licenses and has licensed eight launch sites, which are referred to as spaceports. With the forecasted growth [1] in the number of spaceports and the increase in the number of launches, the challenge now is how to seamlessly integrate the different modes of commercial space activity (Fig. 4.2) into the National Airspace System (NAS).

HORIZONTAL TAKEOFF	VERTICAL LAUNCH
➤ Typically can comply with ATC clearances ➤ Greater horizontal velocity ➤ Large ground footprint ➤ More time to transition through NAS	➤ Typically cannot comply with ATC clearances ➤ Greater vertical velocity ➤ Small ground footprint ➤ Less time to transition through NAS

RE-ENTRY AND RETURN TO LANDING SITE

POWERED FLIGHT	UNPOWERED FLIGHT	BALLISTIC RETURN
➤ Typically can comply with ATC clearances ➤ Discretionary committal to land ➤ Large ground footprint ➤ More time to transition from space	➤ Cannot comply with ATC clearances ➤ Committed to land shortly after de-orbit ➤ Moderate ground footprint ➤ Moderate time to transition from space	➤ Cannot comply with ATC clearances ➤ Committed to land upon de-orbit ➤ Small ground footprint ➤ Minimal time to transition from space

Fig. 4.2 Modes of commercial space activity

To ensure the safety of the public, the AST must somehow segregate launch and re-entry operations from commercial airspace. One way this is achieved is by using aircraft hazard areas or space transition corridors (STCs) (Fig. 4.3) that extend from the ground all the way to space [2–5]. These STCs are sized to ensure that any debris generated by a disintegrating vehicle is contained in an area large enough to pose minimal risk to the public and ground infrastructure.

With the forecast increase in launches [1] over the next few years, these areas of sterilized restricted airspace will potentially create conflict between space operators and NAS users such as commercial airlines. Today, the FAA has no policy that pri-

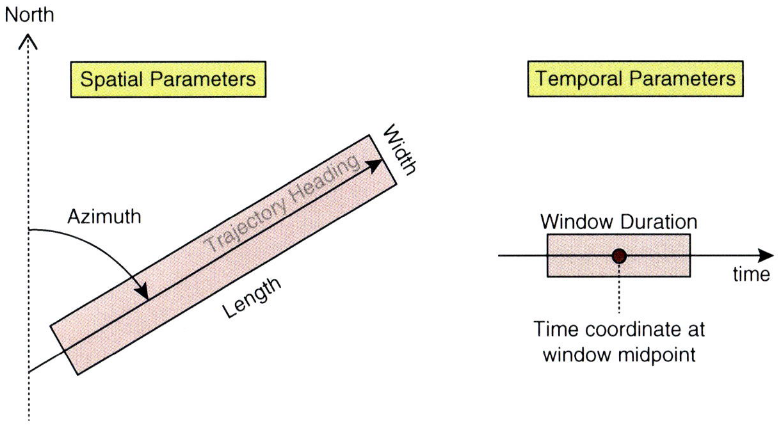

Fig. 4.3 Space transition corridor. Image courtesy of Federal Aviation Administration

oritizes use of the NAS, instead applying a case-by-case and mission-by-mission approach that considers the following:

(i) Minimizing the length of the launch/re-entry window
(ii) Planning the launch window for off-peak times
(iii) Modifying the launch/re-entry trajectory to avoid airspace conflict.
(iv) Negotiating release of airspace with relevant agencies
(v) Creating STCs in aircraft hazard areas.
(vi) Applying a responsive approach that computes real-time hazard areas in a way that permits a tactical response to contingency events.

The approach outlined above was utilized by the FAA for Space Shuttle re-entries following the Columbia accident [6–8] and more recently it has been used for SpaceX's Dragon re-entries. The approach has worked well, but the goal of integrating the growing number of commercial launches remains an ongoing task. To that end, the FAA has designated a Commercial Space Point of Contact (POC) whose job it is to coordinate information between commercial air traffic and space launch operators. The Commercial Space POC is also responsible for distributing mission-specific information to ATC facilities (Fig. 4.4) and NAS users that may be affected by an impending launch. In addition to the Commercial Space POC, the AST has stationed representatives at the Air Traffic Control System Command Center (ATCSCC) to help them understand the process of launching a vehicle into space and also to help the AST devise strategies to overcome the constraints imposed by air traffic management.

Fig. 4.4 Mission control center and air traffic control will need to operate seamlessly when commercial space operations become more frequent. Image courtesy of NASA

FAA Planning Process

Currently, the FAA's approach [9–11] to integrating launch vehicles into the NAS is through effective planning, which comprises the following steps:

(i) **Input data gathering**: in this step, a launch operator provides the FAA with details of the launch, including trajectory, timing and length of the operational window, alternate dates of launch, and any constraints such as weather, wind and/or visibility, orbital dynamics and safety requirements. For example, a suborbital mission studying noctilucent clouds (Fig. 4.5) may require specific atmospheric conditions. Ideally the operator provides this information as soon as possible so feedback can be provided.

(ii) **Plan development and constraint identification**: at this stage, the FAA team assigned to the launch checks information provided by the operator to see if there is conflict with commercial air traffic, aviation events such as air shows, or sports events such as the Daytona 500 that require special airspace management. While the Commercial Space POC works with the ATC facilities that will be affected by the launch to identify any issues and/or constraints, the AST calculate aircraft hazard areas based on information provided by the launch operator.

(iii) **Assessment**: using datasets and planning tools, the Commercial Space POC characterizes the effect of the launch on NAS capacity. The complexity of this

Fig. 4.5 Noctilucent clouds. Image courtesy of NASA

is driven by the type of mission (orbital vs. suborbital), the time of day, and the location. Factors also typically considered in such an assessment include how airspace hazard areas affect the functioning of ATC facilities and the volume and traffic patterns of commercial aircraft transitioning the launch corridor. Using airspace management strategies, the FAA evaluates the launch plan, assesses lessons learned from previous missions, and assesses the impact on NAS efficiency. This information is then passed on to the decision-makers.

(iv) **Decision-making**: the decision to implement a particular strategy is determined by the affected ATC facilities based on constraints and resources.

(v) **Final Planning**: the final step requires the Commercial Space POC to coordinate the selected strategy among the ATC facilities and disseminate the launch information to the representatives of the stakeholders concerned. This may include issuing Notices to Airmen (NOTAMs) for example.

Launch Monitoring

Many of the processes used by the FAA to monitor launch and re-entry operations were developed in response to the Columbia accident [12–14]. Over the years, these processes have been tweaked and refined to enhance situational awareness (SA) and to better respond to contingencies that may impact the NAS. Real-time monitoring is conducted by the ATCSCC which centralizes data flow and communications. Information is also exchanged between the FAA and the vehicle operator so the FAA can track and monitor mission events against the timeline provided to it by the operator. To reduce the impact on the NAS, restrictions are implemented at the last possible moment and are removed as quickly as possible. In the event of a contingency, the FAA communicates with the operator using a hotline, and best estimates of location and extent of hazarded airspace are computed. Following the mission, the FAA reviews the plans that were implemented, evaluates the mission outcome against the plans, and troubleshoots any shortcomings in advance of the next mission. This process is achieved in collaboration with the ATC facilities involved and the vehicle operator.

Integrating Launch Vehicles into the National Airspace System

Suborbital Reusable Launch Vehicles

As of this writing in 2016 there have been eight manned suborbital spaceflights: two X- 15 flights, two Mercury flights, a Soyuz 18a flight and three SpaceShipOne flights. But the next 10–15 years may witness an increasing number of manned suborbital (Fig. 4.6) flights which may ultimately lead to the development of regular commercial suborbital transportation. As part of the long-term plan for these flights it will be necessary to define the risks to civil aviation that may arise by the interaction of suborbital spacecraft and commercial aircraft operating in controlled air space[1]. To do this the NAS will need to accommodate a growing number of horizontal takeoff, horizontal landing (HTHL), vertical takeoff horizontal landing (VTHL), vertical launch vertical landing (VLVL) and air-launched horizontal landing (ALHL) suborbital spacecraft [15].

Flight trajectories for suborbital spacecraft (Fig. 4.7) are very distinct from the flight profiles of conventional commercial aircraft and airspace cannot always be sterilized to accommodate commercial spaceflight operations [14]. This means that spaceports supporting these suborbital spaceflights will need to coordinate launch and re-entry phases with regional ATC and space traffic management (STM) centers to ensure the spacecraft operator and aircraft operator have the necessary SA to safely transition vehicles through air traffic routes [2–4, 10]. To achieve this ATC and STM control centers will need to coordinate the following traffic activities in the NAS (in the scenario depicted below the profile of a HTHL suborbital spacecraft is used):

Fig. 4.6 Launch of New Shepard. Image courtesy of Blue Origin

Launch/Takeoff

Pre-departure actions for a HTHL will be as follows:

i. Once a mission plan is been filed, TM previews the plan. Sequencing and scheduling tools are used to provide TM advisories and ensure the launch is coordinated with arrival and departure traffic.
ii. ATC controls departure activities, establishes the necessary communication links, and reviews the necessary TM initiatives required, such as STCs and temporary routes for aircraft.
iii. Departure status is displayed to ATC and MCC controllers. ATC monitors STC status and separates traffic from active STCs. A conflict prediction tool assists ATC in detecting aircraft that conflicts with the STC.

Ascent

Entry to suborbital space requires a nearly vertical, high-acceleration ascent phase that is under control of ATC to 60,000 ft (Flight Level 600, or FL600): at this altitude, control is handed over to MCC. The HTHL vehicle makes a conventional transition through the NAS. The following paragraphs describe the phases of such an ascent [10].

ATC from the Surface to a High-Altitude STC

After being cleared the vehicle is protected by a STC to the upper limit (normally 60,000 ft) of the NAS. For ATC to be used, the vehicle must be able to be tracked and the MCC must have direct voice/datalink communications with ATC. This mode is only used when traffic and environmental conditions yield acceptable levels of controller workload.

ATC During Initial Ascent

Upon departure, vehicle position is tracked on ATC displays and ATC issues clearances to the MCC to assure separation for the vehicle while it transitions to the point at which the vertical ascent is initiated. Conflict prediction software recognizes the vehicle's performance envelope and hazard volume and provides appropriate resolutions. Collision avoidance systems used by aircraft recognize the vehicle and its performance profile to provide conflict advisories.

Final Ascent Through a STC

As the vehicle approaches the point for vertical ascent, a high altitude STC is activated. The STC is tailored to vehicle characteristics (speed, rate of acceleration, trajectory). The STC data is depicted on TM/controller displays, along with weather and traffic data. Controllers of the sectors that contain the STC communicate with the MCC to ensure the STC is activated and clear of traffic. Conflict prediction and resolution advisories provide notification of flights that are predicted to conflict with the STC [10].

Ascent Through STC from the Surface

Upon departure, vehicles not controlled by ATC may be cleared on a flexible spaceway. In this case the vehicle's ascent is conducted as follows:

(1) The vehicle's position and STC status is available on situation displays at ATC and TM. STC strata is reserved and released as the vehicle progresses through its flight trajectory while STC schedule, status, and coordinates are depicted on integrated TM and controller displays: this data is complemented by weather and traffic information to provide a comprehensive view of the flight.
(2) The sectors containing the STC communicate with TM to ensure the STC is clear of traffic. To ease the burden of monitoring the STC, controllers receive event prompts for STC status changes and conflict prediction/resolution advisories provide notification of flights that conflict with the STC. Using DSS, advisories

and advanced situation displays, ATC anticipate future airspace status and direct flights through/away from the airspace based on availability.
(3) STC information is available on the flight decks of aircraft equipped with traffic displays.
(4) The MCC informs ATC when the vehicle exits the NAS, and the STC is released.

Re-Entry

Throughout the mission, status information may be accessed by NAS users for strategic planning. If the re-entry plan must be modified, the MCC coordinates the revised plan with the AOCs. The re-entry is conducted as follows:

(1) Re-entry plans are used by TM to define appropriate STCs. Like the STCs used for transitions to space, baseline STCs exist to support typical vehicle re-entries from various points in space. These STCs are dynamically reserved/released, and may be tailored to satisfy specific mission/vehicle profiles.
(2) MCC performs fast-time scheduling analyses to determine the re-entry plan's feasibility, and coordinates with the appropriate sectors. ATC issues the re-entry clearance. For unpowered returns, re-entry clearance is an implied clearance to land, since the vehicle is committed to its return upon de-orbiting.
(3) The MCC disseminates notification to AOCs via the NAS-WIS. Affected sectors are notified by MCC of the re-entry and STC activation and STC event prompts assist in establishing SA of the traffic situation. Conflict prediction and resolution capabilities help controllers mitigate conflicts between air traffic and active STCs, and to resume use of the airspace as STC strata are released, while sequencing and scheduling advisories assist in managing traffic flows.
(4) When the vehicle re-enters the NAS, the MCC ensures the vehicle's trajectory conforms to the designated STC. The vehicle's position and STC status are available on situation displays at ATC and TM positions, at the MCC.

Descent Through the NAS and Landing

Descent through the NAS is managed by either:

(1) The vehicle being protected by an STC for the entire transition from the upper limit of the NAS to the surface, or:
(2) The vehicle being protected by an STC from the upper limit of the NAS to a point at which it assumes the performance characteristics of a conventional aircraft, and ATC techniques can be used to monitor the vehicle back to the runway/pad. The application of one of these options to a mission is largely determined by the re-entry profile of the vehicle. Since the HTHL vehicle is a glider upon re-entry, the re-entry follows a gliding return as follows:

Positive ATC is not an option for a gliding return since the vehicle is not capable of responding to the full range of ATC clearances. Upon re-entry, these vehicles have a higher descent rate than powered vehicles, and they must anticipate any constraints to landing since alternative trajectories must be exercised early in the re-entry. It is expected that trajectory modeling will be used to identify the point at which the spacecraft will be committed to landing at the spaceport. This point represents the final opportunity to invoke any contingency plan to route the spacecraft to an alternate landing site if operational conditions are undesirable at the primary site. The MCC and TM maintain direct communications in case contingency plans have to be implemented. When the spacecraft reaches its commitment point, it is handled as a priority vehicle since it does not have the option of deviating from its landing plan.

Coordinating Traffic and Operational Variables

Coordinating traffic (Fig. 4.7) within the NAS will required the following actions [10, 11]:

1. Issuing restrictions
2. Re-routing air traffic
3. De-conflicting traffic
4. Instructing a spaceport to delay a launch
5. Directing suborbital vehicles in the event of an emergency
6. Modifying flight paths/spacecraft transitional corridors due to weather.

This level of coordination will require the development of new traffic management (TM) procedures that will need to be seamlessly integrated into the NAS [11]. These procedures will also need to be dynamically reconfigurable to maximize operational flexibility and to accommodate contingencies. As mentioned earlier, currently space vehicle operations within the NAS are managed using STCs and Flexible Spaceways [11] but little has been done to coordinate the implementation of these traffic flow measures into the NAS. To do this it may be necessary to develop a deci-

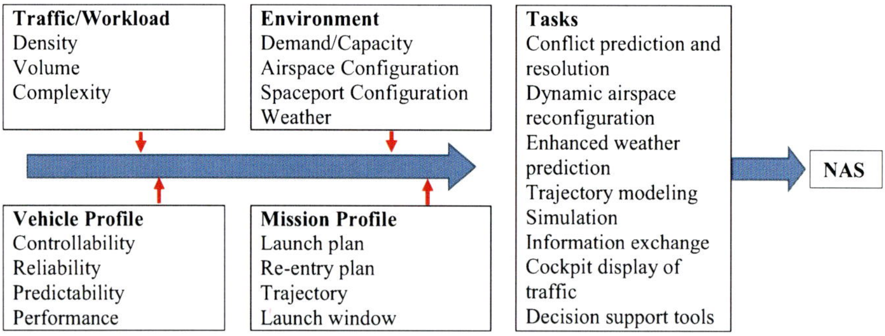

Fig. 4.7 Coordination of NAS activities. Image by the author

sion support system (DSS) based on mission planning for each simulated mission flown, to include the following:

a. Trajectory modeling (Fig. 4.8)
b. Traffic de-confliction and resolution

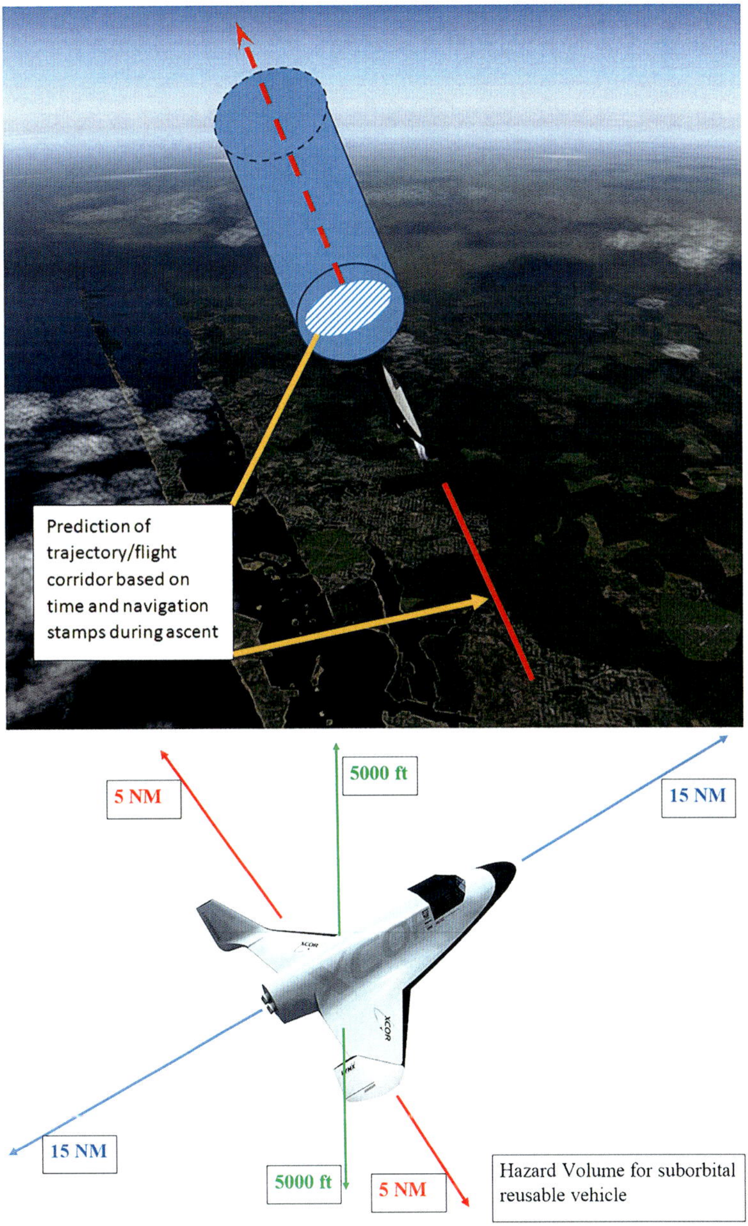

Fig. 4.8 *Upper image*: using predictive algorithms it will be possible for ATC/MCC to calculate trajectories of spacecraft based on inputs such as acceleration, wind shear, and dynamic pressure. *Lower image*: the hazard volume of a suborbital spacecraft. Images by the author

c. Trial plan trajectories
d. Airspace contention
e. Traffic workload prediction
f. Event prompts for ATC and MCC controllers.

Potentially, much of this work could be conducted by 'flying' missions in high fidelity simulators. Simulated/and or real mission operational variables such those detailed below could then be analyzed to integrate space and aviation operations. Potentially, some of this work could be performed at spaceports if simulators are available.

i. Traffic variables such as demand/capacity, dynamic density, volume, and complexity will drive controller workload and define the resources available for handling the mission [10].
ii. Sector, airport/spaceport configuration, and weather conditions will influence controller workload [15].
iii. The vehicle profile will determine traffic management options. Determinations of these options will be based on the vehicle's ability to comply with ATC clearances.
iv. The mission profile will determine the impact of launch/re-entry on traffic flow through the NAS. The impact of the mission on the NAS is determined by:

    a. Departure location
    b. Trajectory, re-entry location
    c. Trajectory, landing location
    d. Launch/re-entry window sizes
    e. Instantaneous impact points (IIPs)
    f. Vehicles with hazardous payloads require increased buffer zones, and/or trajectories that minimize the ground footprint over populated areas. The combination of the mission/vehicle profile will dictate options available in case of an aborted mission.

Factors other than launch and re-entry technique will affect a vehicle's ability to comply with ATC clearances. These factors include:

i. *Vehicle performance*: Vehicles that perform comparably to conventional aircraft *may* be controlled via ATC techniques whereas vehicles with ballistic profiles will require reserved airspace to ensure separation during transition to/from space.
ii. *Vehicle equipage*: The electronic systems used by the vehicle will determine the vehicle's ability to respond to ATC and the precision with which the vehicle can be tracked.
iii. *Trajectory predictability*: Vehicle performance and reliability, mode of pilotage, and tracking capability will determine NAS accuracy of trajectory predictions for traffic planning and deconfliction. For example, trajectories may be affected by wind shear and aircraft may need to be diverted due to a launch vehicle contingency or a problem encountered during a gliding re-entry.
iv. *Trajectory Modeling/Simulation Tools*: Predictive simulation tools will enable traffic managers and mission planners to anticipate conflicts between space

traffic and air traffic, including conflicts with active STCs (4). For example, how will commercial air traffic be rerouted in the event of a conflict with a returning spacecraft, or a spacecraft that has suffered an 'engine out' failure?

v. *Conflict Prediction & Resolution*: Automated trajectory modeling will assist mission planners to predict and resolve conflicts between spacecraft, aircraft, airspace, and weather.

Flight Testing Overview

As can be appreciated by looking at all these factors, the process of integrating vehicles into the NAS will be a difficult one, especially since launches are relatively few and far between. But when launches become more commonplace, the database of launches will be able to develop DSS tools based on factors such as:

i. Payload/Manifest (e.g., passengers, hazardous cargo)
ii. Mission duration
iii. Vehicle type
iv. Launch window
v. Flight profile/Preferred route/trajectory (launch & re-entry)
vi. Re-entry window (primary, secondary)
vii. Launch location
viii. Estimated Trajectory/Initial heading & azimuth
ix. Air Defense Identification Zones (ADIS) penetrated
x. Point of re-entry
xi. STC information
xii. Landing location
xiii. IIP [16]
xiv. Traffic loading at time of launch & re-entry
xv. Weather (current & forecast)
xvi. Airspace configuration (e.g., active MOAs, etc.)

After each flight, vehicle performance, vehicle trajectory, traffic loading, dynamic density, and weather and atmospheric conditions will be analyzed to measure the interactions between these variables. Airspace requirements, mission scenarios, and availability of NAS resources will also be evaluated. On completing post-mission analysis, a series of candidate mission profiles will be developed.

Next Steps

Over the years, as commercial launch operations evolves, the processes and procedures used to integrate launch vehicles into the NAS will become increasingly standardized at spaceports around the world. Today, those procedures and processes focus heavily on protecting from failure, but gradually, as launch operations become more streamlined, this emphasis will move more towards operating for success. At this level, NAS will only be sterilized in response to a vehicle failure, but such a level of operation will require a lot of investment on the part of vehicle operators to increase vehicle reliability. Such a level of operation will only be achieved through high frequency of operations and the development of spaceport-specific technologies that enhance operational predictability. In tandem with the improvements made by the operator, the FAA will need to invest in techniques that enhance operational predictability and ATC/spaceport mission control type capabilities.

NAS management is capable of overcoming severe dynamic weather, military exercises, and variable system capacity demands. To ensure the seamless integration of sRLVs into the NAS, high data rate processing, flight planning and real time data communication will be required. Due to the high speeds of sRLVs, dynamic surveillance and tracking data must be processed by the NAS, while being displayed alongside aircraft traffic data. Flight planning for sRLVs will be required in advance of missions, and much earlier than commercial aviation flight plans. These plans will allow ATC-MCC to manage the mission impacts to the NAS, while adapting to any sRLV off nominal events through planning and prediction tools that will be generated by conducting simulation.

References

1. The Tauri Group: Suborbital Reusable Vehicles: A 10-Year Forecast of Market Demand 2–24, (2012)
2. Murray, D.P.: The FAA's Current Approach to Integrating Commercial Space Operations into the National Airspace System. Federal Aviation Administration, Washington, DC (2013)
3. Murray, D.: Challenges in Accommodating Commercial Space Operations in the NAS. Office of Commercial Transportation, Federal Aviation Administration, Washington, DC (2014)
4. Bilimoria, K.D., Jastrzebski, M.: Space transition corridors in the National Airspace System. Aviation Technology, Integration, and Operations Conference, Los Angeles, CA, August 12–14, 2013
5. Gonzales, E., Murray, D.: FAA's approach to ground and NAS separation distances for commercial rocket launches. Proceedings of the 48th AIAA Aerospace Sciences Meeting, 2010
6. Johnson, C.W.: The future of sub-orbital and orbital space accident investigation. In: Baram, M. (ed.), Proceedings of the New Technologies and Work Network, Toulouse, France, 11–13th October 2012, Network Consortium, France, 2012. Available on: http://www.dcs.gla.ac.uk/~johnson/papers/Space/SpaceAccidents2.pdf
7. Zapata, E: Separation distances for rocket launch operations. Proceedings of the AIAA Atmospheric Flight Mechanics Conference and Exhibit, August 2008
8. Mrozinski, R: Space shuttle probabilistic risk assessment incorporation into entry public risk estimates. Proceedings of the AIAA Atmospheric Flight Mechanics Conference and Exhibit, August 2005
9. Vansuetendael, R., Hayes, A., Birr, R.: A common communications, navigation & surveillance infrastructure for accommodating space vehicles in the next generation air transportation system. AIAA Guidance, Navigation and Control Conference and Exhibit, AIAA Atmospheric Flight Mechanics Conference and Exhibit, AIAA Modeling and Simulation Technologies Conference and Exhibit, AIAA Paper 2008–6893 (2008)

10. Murray, D., Van Suetendael, R.L.: A tool for integrating commercial space operations into the National Airspace System. In: Proceedings of the American Institute of Aeronautics and Astronautics Aerospace Atmospheric Flight Mechanics Conference and Exhibit, Keystone, Colorado, 21–24 August 2006
11. Murray, D., Van Suetendael, R.: A tool for integrating commercial space operations into the National Airspace System. Proceedings of the AIAA Atmospheric Flight Mechanics Conference and Exhibit, August 2006
12. Columbia Accident Investigation Board (CAIB): CAIB Report, Washington, D.C., August 2003, Vol. I.
13. Federal Aviation Administration: Guide to Probability of Failure Analysis for New Expendable Launch Vehicles, November 2005. http://www.faa.gov/about/office_org/headquarters_offices/ast/licenses_permits/media/Guide_Probability_Failure_110205.pdf
14. Federal Aviation Administration: Guide to Reusable Launch and Reentry Vehicle Reliability Analysis, April 2005. http://www.faa.gov/about/office_org/headquarters_offices/ast/licenses_permits/media/FAA_AST_Guide_to_Reliability_Analysis_v1.pdf
15. Duan, P.: ADS-B feasibility study for commercial space flight operations. IEEE Xplore, November 2010
16. Larson, E., Carbon, S., Murray, D.: Automated calculation of aircraft hazard areas from space vehicle accidents: application to the shuttle. Proceedings of the AIAA Atmospheric Flight Mechanics Conference and Exhibit, August 2008

Chapter 5
Spacecraft and Launch Vehicles

Fig. 5.1 Launch vehicle transportation facilities are just one of the myriad services that must be offered by today's spaceport. Image courtesy of NASA

The term 'spaceport' encompasses a broad variety of facilities (Fig. 5.1). Some spaceports are used to launch large rockets into Earth orbit, some of which carry their payloads into deep space while others carry probes and yet others ferry cargo and/or astronauts. Then there are spaceports that have been developed for launching jet-like spacecraft from runways for trips into suborbital space. And at the most basic level there are the spaceports that simply support short flights with small

E. Seedhouse, *Spaceports Around the World, A Global Growth Industry*,
SpringerBriefs in Space Development, DOI 10.1007/978-3-319-46846-4_5

rockets. But one factor common to all spaceports is that each is the location where space launch vehicles and their payloads are prepared and subsequently launched. As of October 2016 there were eight commercially licensed spaceports in the United States, but the regulations that governed the application process (14 CFR Part 420) were developed at a time when most launch operators were in the business of launching orbital launch vehicles that operated from federal launch ranges. But since those regulations came into effect the launch vehicle arena has changed significantly. Perhaps the most notable change has been with the arrival of suborbital reusable launch vehicles (sRLV) and the development of private launch sites such as Blue Origin's remote location in Van Horn, Texas. Since spaceport design is driven largely by the types of launch vehicles being launched, it is useful to know what capabilities these vehicles have and the facilities required to support their launch and recovery. To that end, what follows is a synopsis of some of the current crop of orbital and suborbital vehicles around the world.

Orbital Vehicles

United States

In the U.S. it is unlikely there will be dramatic changes in the types of small to heavy orbital launch vehicles until the mid 2020s. The current crop of expendable launch vehicles (ELVs) such as the Atlas V (Fig. 5.2) and the Delta IV have proven reliable and large enough to meet current and forecasted needs for the next several years at least. In addition to United Launch Alliance's (ULA) Atlas and Delta vehicles there is also the Falcon (Fig. 5.3) family of launch vehicles that are launched by SpaceX to deliver Dragons (cargo and, beginning in 2018, astronauts) to the International Space Station (ISS). SpaceX also launches satellites from Vandenberg Air Force Base (VAFB) using its Falcon Heavy.

Among the new generation of ELVs' is the Space Launch System (SLS) which is being developed to support Exploration Class missions in the late 2020s and early 2030s. The SLS (Fig. 5.4), which NASA expects to be operational in the early 2020s, will be capable of delivering 70 t to low Earth orbit (LEO). A larger variant will be capable of launching 130 t, but this vehicle is unlikely to be operational until the late 2020s.

In parallel with the development of the SLS is the development of launch vehicles that are reusable, some of which may make an appearance before the end of the 2010s. One such vehicle is Blue Origin's New Shepard vehicle, a manned suborbital spacecraft that will eventually be developed into a manned orbital vehicle to be launched from Florida's spaceport. Other manned vehicles include SpaceX's Dragon V2 (Fig. 5.5) which will ferry crews to the ISS beginning in 2018, and Boeing's CST-100 (aka the Starliner) spacecraft which will do the same in 2018. Another crew vehicle being developed for ISS duty is NASA's costly Orion Multi-Purpose Crew Vehicle (MPCV) that will launch from Florida sometime before the end of the 2010s.

Fig. 5.2 A United Launch Alliance Atlas 5 rocket blasts off from Space Launch Complex-41 Station on February 24, 2012, with the U.S. Navy's MUOS-1 satellite. Image courtesy of USN

Fig. 5.3 SpaceX Falcon 9 rocket completes static fire test on December 19, 2014. Image adapted from *SpaceX: Making Commercial Spaceflight a Reality*

Fig. 5.4 NASA's next generation launch system: Space Launch System. Image courtesy of NASA

Fig. 5.5 SpaceX's Dragon V2. Image courtesy of NASA

Russia

Russia has the most reliable launch vehicle in the world and its name is Soyuz (Fig. 5.6). With an impressive resume of more than 1700 unmanned and manned launches, the Soyuz vehicle is by far the most used rocket on the planet with a legacy stretching back to 1967. Used to launch crew to the ISS and to launch unmanned Progress supply craft to the orbiting outpost, the Soyuz is also used for commercial launches marketed by ArianeSpace.

Despite being around seemingly forever, the Soyuz has suffered surprisingly few mishaps, although a clutch of incidents occurred in the early 2000s and 2010s. First there was the loss of the unmanned Photon-M satellite in October 2002 that resulted in an explosion killing one ground crew. This was followed 3 years later by the loss of a Molniya satellite. Then, in 2011, a Soyuz-U carrying supplies to the ISS crashed, an event followed just months later by a failure of a Soyuz-2.1b that was launching a Meridian-5 communication satellite. But the venerable launcher continues to be the mainstay of the ISS program and a popular choice for companies launching commercial satellites. Since 2011 the Soyuz has been launched from the European Space Agency's (ESA) Guyana Space Center in Kourou.

Another dependable Russian launch system is the Proton, which is used to launch commercial and government payloads. Like the Soyuz, the Proton has a history stretching back decades, to 1965, when the first rocket was launched. 50 years later

Fig. 5.6 Soyuz. Image courtesy of NASA

and the vehicle is still being used to ferry supplies to the ISS. In its original designation the Proton was a super heavy ICBM, capable of launching a 100 megaton nuclear weapon more than 10,000 km, but nowadays its capability is 22.8 t to LEO. The rocket is a dependable workhorse that has seen few failures: most recent failures include the 2012 loss of a Russian and Indonesian communications satellite and the 2013 loss of three navigation satellites. Over the years the Proton has had minor upgrades to its software, flight control, and aerodynamics, but major upgrades were put on ice due to the development of the new Angara rocket.

The Angara (Fig. 5.7) medium-lift rocket first launched from the Plesetsk Cosmodrome on July 9. 2014. It was a notable event since the Angara was the first vehicle to be built from the ground up since the demise of the Soviet Union. When complete, the Angara family of launch vehicles will include light, medium and heavy-lift variants built on a modular principle that uses generic boosters. Western observers suspect the Angara will eventually replace the Proton which is slated for retirement around 2030.

China

China announced its entry into the business of launching rockets in 1970 with the launch of its first satellite, the Dong Fang Hong 1, which was carried aloft a Long March rocket. More than four decades later, the Long March family continues to be

Fig. 5.7 Angara Rocket. Image courtesy of Russian Space Agency

the lynchpin of the country's launch capability. Since Dong Fang Hong 1, the Chinese have used the Long March vehicles to launch the country's first taikonaut and the Chang'e lunar orbiters. As of this writing, the Long March tradition continues and there have been 226 Long March (Fig. 5.8) missions.

Fig. 5.8 Long March 4B. Image courtesy of Pline

In June 2016 China launched its 53-m tall Long March 7 rocket, the latest in a long line of Long Marches, and the debut of a vehicle capable of lifting more than 13 t into LEO. The launch also marked the opening of the Wenchang Satellite Launch Centre, located at China's most southerly point, which means it will be easier for China to launch satellites into geostationary orbits.

India

India's first satellite was launched by the Soviet Union in 1975 but it wasn't until 1980 that the country launched its first satellite on an Indian-made vehicle. Since then, the Indian Space Research Organisation has developed a family of launch vehicles as follows:

- Polar Satellite Launch Vehicle—a variant of this—the PSLV XL—launched India's Chandrayaan-1 lunar orbiter in 2008. This launch vehicle also launched the Mars Orbiter Mission in 2013. As of 2016 this vehicle (Fig. 5.9) has launched more than 70 satellites both Indian and foreign.
- Geosynchronous Satellite Launch Vehicle: India's second heaviest satellite launch vehicle, capable of lifting 5 t into LEO.
- Satellite Launch Vehicle: this vehicle was retired in 1983
- Augmented Satellite Launch Vehicle: first launched in 1987, this launch vehicle was decommissioned in 1994.
- Geosynchronous Satellite Launch Vehicle Mark-III. As of 2016, this vehicle is still being developed. When operational it will have a lift-off mass of 640 t and be capable of launching up to 4 t into geosynchronous orbit.

Suborbital Vehicles

There has been a lot of talk about launching suborbital vehicles into space since SpaceShipOne (SS1) won the X-Prize in 2004, but since that aviation milestone there hasn't been a single manned suborbital flight. In 2014 the commercial space industry expected SpaceShipTwo (SS2) to make its maiden suborbital flight from the Mojave Air and Space Port, but the deadly accident in October that year deep-sixed that event. The truth is, the Mojave Air and Space Port, which used to be a ground zero for the development of manned suborbital vehicles, sees very little traffic of any kind, and after the Lynx (Fig. 5.9) was moth-balled in June 2016, the chances of routine manned suborbital flights before the end of the decade are slim to none. Meanwhile, New Mexico officials are biting their nails waiting for the completion of Virgin Galactic's SS2 Mk II flight test program. Once this is completed, the state's spaceport—Spaceport America—will finally be able to welcome its anchor tenant and its spacecraft.

Fig. 5.9 The Lynx spacecraft which was mothballed in June 2016. Image adapted from *XCOR, Developing the Next Generation Spaceplane*

Fig. 5.10 The troubled SpaceShipTwo. Image courtesy of Virgin Galactic

SpaceShipTwo

On 29 April, 2013, SS2, a spaceship financed by Sir Richard Branson, made its first powered flight over Mojave, California. Although SS2 (Fig. 5.10) didn't actually fly in space during the test flight, it marked a significant milestone for the company that had spent the best part of a decade developing the successor to SS1. During the April early morning flight, SS2, strapped beneath its mother-ship, WhiteKnightTwo (WK2), took off from a runway in the Mojave Desert. Once it reached the release altitude, WK2 released SS2, and the compact spacecraft ignited its engine for 16 s, before gliding to a safe landing. Although only 16 s of the 13-min flight were powered, the test moved Virgin Galactic one significant step closer towards its goal of flying passengers into space. Coming less than 3 years after SS2's first glide flight, the powered test marked the beginning of the envelope expansion phase and the very real possibility of flights into space by the end of 2013. But it wasn't to be.

Manufactured by The Spaceship Company, SS2 is a spacecraft designed to ferry up to six passengers into space at a cost of $250,000 per ticket. In the first few years following SS1's historic flight, Virgin Galactic announced a series of development and construction milestones, but momentum ground to a halt in 2007 when a fatal fire during a ground test killed three. The California Occupational Safety and Health Administration investigated the accident, noting that Scaled Composites had failed to provide adequate training about the hazards involved with the nitrous oxide rocket fuel the company used in its spacecraft prior to the accident. The California State investigation found Scaled guilty of failing to observe correct workplace practices, but was unable to explain what had happened. Work stopped on SS2 for a year and once again Virgin Galactic had to revise its forecast for revenue flights. The estimated costs of the program, first calculated at $20 million, rose to $400 million. Eighteen months later, Virgin Galactic unveiled the new SS2 (sidebar) to a crowd of 800, press, future astronauts and VIP guests, including Governors Bill Richardson and Arnold Schwarzenegger. After being carried down the runway by her mother-ship, VMS Eve, to a spectacular display of floodlights and music, SS1's successor was christened VSS Enterprise, after its Star Trek namesake.

SS2 Specifications

Crew: 2
Passengers: 6
Length: 18.3 m
Wingspan: 8.3 m
Height (rudders down): 5.5 m
Cabin diameter: 2.3 m

Flight Profile

1. SS2 carried to 15,500 m by WK2.
2. After release from WK2 SS2 fires its rocket engine for 70 s and accelerates to 4000 km per hour.
3. After the rocket engine is shut down SS2 enters the coast phase which carries it to an altitude higher than 100,000 m.
4. Passengers experience up to 5 min of weightlessness.
5. Re-entry with rudders in feathered configuration.
6. At 22,900 m rudders are defeathered and SS2 becomes a glider.
7. Landing gear is deployed and SS2 lands on runway.

Before SS2 can start taking off from spaceports, one of the safety issues will be how the Federal Aviation Administration (FAA) certify SS2 safe for passenger flight. In the commercial airline world it may take 2 years for the FAA to clear a new airliner for service, and this process involves extremely well understood technical issues and mature technology. In short, when it comes to your 787's, there is little that has never been done before, and most that hasn't been done before involves software, *not* structure. The challenge faced by Virgin Galactic is that *nothing* has been done before. In the wake of SpaceShipOne's success, the U.S. Congress debated how to regulate commercial human spaceflight, arguing at length about how to handle crew and passenger safety. The end result of all the debating was the Commercial Space Launch Amendments Act (CSLAA), which was signed into law by President George W. Bush on December 23, 2004. That bill included a provision that restricted the FAA's Office of Commercial Space Transportation (AST) from enacting safety regulations except for cases linked to the "serious or fatal injury" of crew or participants, or events that "posed a high risk" of such injuries, during licensed or permitted flights. According to the law the restriction expired 8 years after the law's enactment, to allow the industry to build experience upon which future safety regulations could be based. But the industry developed far more slowly than anticipated in late 2004. This led to calls for an extension to the CSLAA, which is what happened in late 2013 when the U.S. House of Representatives approved the Space Launch Liability Indemnification Extension Act (H.R. 3547) that extended for 1 year the commercial space transportation risk-sharing and liability regime that was established with passage of the CSLAA. But the CSLAA paperwork didn't allow Virgin Galactic to launch passengers into space because by June 2014 the FAA had yet to grant Virgin Galactic a commercial operator's licence. What's more, the FAA hadn't laid out safety rules like those for aircraft, and there were no plans to do so until at least October 2015. The operator's licence, known as a Reusable Launch Vehicle Mission Licence (sidebar), was the final piece of the commercial space travel puzzle that Virgin Galactic needed before revenue flights could begin. Realizing the licence was one of the company's remaining major milestones, Virgin Galactic submitted an application to the FAA's AST in August 2013. The office had 6 months to review the application, meaning Virgin Galactic could have reasonably expected approval by as early as February 2014 (sidebar).

Committee Passes Bill Allowing FAA to Issue Permits and Licenses for Same Vehicle

In April, 2014, The Senate Committee on Commerce, Science, and Transportation approved S. 2140, legislation that would fix burdensome federal laws to allow for more advancement in the commercial spaceflight industry. The legislation changed current laws that are slowing progress for the reusable launch vehicle industry to ensure space companies can continue to test their vehicles. The bill would allow a commercial space company to take a licensed vehicle out of commercial service and use it as an experimental platform for safety and performance improvements when needed, and allow one or more vehicles of the same design to be used for test flights under a permit, while other vehicles of the same design are used in commercial operations under a license. For Virgin Galactic, the bill was a step in the right direction, by not only addressing key technical issues, but also updating and streamlining federal laws so they are more in line with today's commercial spaceflight operations.

§431.3 Types of Reusable Launch Vehicle Mission Licenses
(a) *Mission-specific license.* A mission-specific license authorizing an RLV mission authorizes a licensee to launch and reenter, or otherwise land, one model or type of RLV from a launch site approved for the mission to a re-entry site or other location approved for the mission. A mission-specific license authorizing an RLV mission may authorize more than one RLV mission and identifies each flight of an RLV authorized under the license. A licensee's authorization to conduct RLV missions terminates upon completion of all activities authorized by the license or the expiration date stated in the re-entry license, whichever occurs first.

(b) *Operator license.* An operator license for RLV missions authorizes a licensee to launch and reenter, or otherwise land, any of a designated family of RLVs within authorized parameters, including launch sites and trajectories, transporting specified classes of payloads to any re-entry site or other location designated in the license. An operator license for RLV missions is valid for a 2-year renewable term.

But February 2014 came and went and Virgin Galactic still didn't have their licence. Why? Well, in 2014 one of the challenges facing regulators drafting safety rules was the lack of test flights (sidebar) that had been conducted. When the FAA created its process for regulating the industry it was anticipated that dozens if not hundreds of test flights would have occurred by 2014, providing plenty of data that could have been used to inform the drafting of rules, but that didn't happen. And, until SS2 actually reaches its target performance with six passengers aboard, nobody knows how it will perform.

Society's Approach to Risk
According to the National Safety Council, an American has about a 1-in-80 chance of dying in a car accident in their lifetime versus a 1-in-4608 for a flying accident. But how many people have a fear of driving? It's all about risk perception. Now think back to the Columbia accident. NASA officials and some members of Congress argued that NASA needed to get back to flying the Shuttle as soon as the fixes recommended by the Columbia Accident Investigation Board were implemented. The public agreed, with 68 % (according to Zogby) saying they thought the benefits of manned spaceflight outweighed the benefits. But, with an accident rate of one every 62.5 missions, in which 14 astronauts lost their lives, translating that risk to commercial aviation would mean thousands of people would be killed *every day*. So what level of risk is acceptable for those flying Virgin Galactic? 1 %? 0.1 %? 0.01 %? And who makes the call?

By the end of May 2014 Virgin Galactic had secured the rights to plan the world's first space tourism flights. That meant Branson had secured permission from the FAA to begin setting guidelines and plans in motion before the actual launches begin. Whilst the agreement with the FAA didn't give Virgin Galactic permission to begin launches, it was another step towards revenue flights because it meant Virgin Galactic was authorized to operate like a regular commercial airline, and would be included in the United States National Airspace System (NAS). Then, in October 2014, SS2 crashed and the program was stalled yet again. The end of 2016 marked a return to flight of sorts with Virgin Galactic conducting its first glide tests since 2014. Now all Virgin Galactic has to do is make sure SS2 Mk II passes the testing and safety procedures required by law to enter space commercially—as will Blue Origin, which is in the same business as Virgin Galactic.

Blue Origin

Founded by Jeff Bezos of Amazon.com fame, Blue Origin is developing its vertical takeoff, vertical landing (VTVL) spaceship New Shepard, to fly suborbital and orbital trips. New Shepard's test flights have gone very smoothly, with the fourth taking place on 19 June 2016. Manned test flights are expected to take place in late 2017 and revenue flights could start as early as 2019, which will place the company way ahead of Virgin Galactic. Unlike Virgin Galactic, Blue Origin plans to fly its suborbital flights from its Culberson County, Texas facility, which hardly qualifies as a spaceport, although orbital services will launch from the Florida Space Coast.

Chapter 6
Payload Processing, Testing and Integration

Fig. 6.1 Arianespace's payload processing bay. Image courtesy of European Space Agency

Payload processing is the business of preparing, integrating, testing and launching payloads into space (Fig. 6.1). Given the projected demand for passenger and science missions, spaceports must be prepared to offer the variety of services, equipment and facilities required to prepare payloads for their journey into space. This chapter provides an overview of the key elements involved in the payload processing sequence and the facilities required to support these activities.

E. Seedhouse, *Spaceports Around the World, A Global Growth Industry*,
SpringerBriefs in Space Development, DOI 10.1007/978-3-319-46846-4_6

Payload Integration Administration

Before the payload arrives at the spaceport there is a fair amount of paperwork that must be completed. These tasks are dealt with by the payload integration manager who serves as the principal investigator's primary point of contact for all matters pertaining to the payload. The payload integration manager is responsible for developing the payload Interface Control Document (ICD), the payload schedule and any payload data products. They also manage the data deliveries in support of the mission and ensure payload requirements are correctly defined, documented and compatible with the vehicle. If changes to the payload are required, these changes are noted in the Change Evaluation Form (CEF). Sometime in this phase a Hardware Feasibility Assessment (HFA) is conducted to ensure payload viability. The next stage is a flight readiness review (FRR) to ensure the payload is safe to be flown. If it is, a Certificate of Flight Readiness (CFR) is issued. Once the principal investigator has the CFR they arrange for the payload to be delivered to the spaceport. Once there, a crew equipment interface test may be performed, along with verification of hardware interfaces, functional testing of the payload and hands-on internal and external verification of the payload in flight configuration.

Payload Testing and Integration Step by Step

Many readers may be unfamiliar with the payload test and integration process, so what follows is a step-by-step primer which steers clear of as much technical jargon as possible.

Before a payload can be launched the spaceport must support the customer with a great deal of administrative assistance (Fig. 6.2) that includes assistance with launch site policies, the planning and scheduling of safety plans and the logistical administrative support for the transportation, shipping and receiving of the payload. In addition to these functions, the spaceport must provide the following:

(i) Non-destructive evaluation and testing.
(ii) Means of dealing with hazardous waste.
(iii) Means of evaluating payloads for impact on spaceport facilities and equipment.
(iv) Chemical sampling and analysis.
(v) Assistance on the repair of ground equipment supporting the payload.
(vi) Means of assessing the effect of payloads on ground operations/spaceport personnel.
(vii) Simulation and modelling capabilities to define flight and ground hardware access during payload processing.
(viii) Capabilities for flight hardware integration, to include fit checks, handling, alignment, centre of gravity calculation, servicing and maintenance.

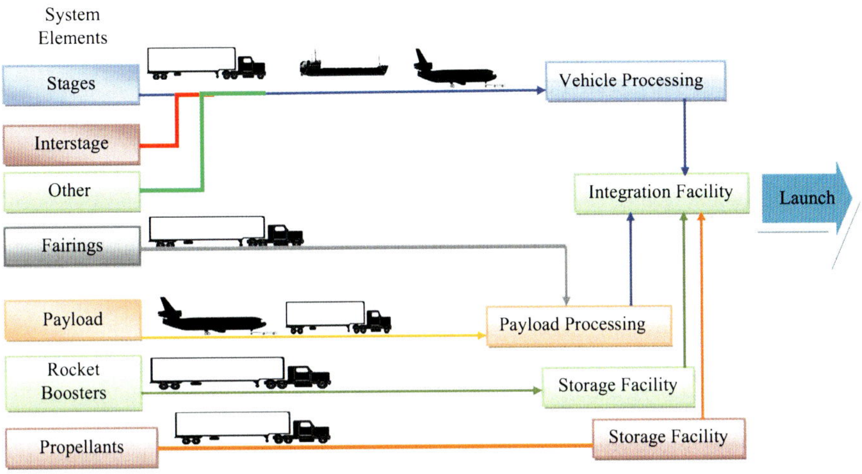

Fig. 6.2 Generic launch vehicle and payload processing overview. Image by the author

Payload Preparation and Integration Phase by Phase

The first step in the process of preparing a payload for launch is the Strategic Phase (Fig. 6.3), which includes the payload requirements, the payload design, safety reviews and the flight assignment, which is when the payload makes its way onto the spacecraft manifest. The next phase is the Tactical Phase which includes crew training, which for suborbital flights will usually include the Principal Investigator and a back-up. This is followed by the Operations Phase, which encompasses the flight of the payload while the Post-flight Phase includes de-integration, the return of the payload from the landing site, lessons learned, data return/processing, crew debrief, and the mission report. These steps are described in Fig. 6.3 below. Within the process is an integration flow which includes the following:

- Payload Performance
- Payload Manifest
- Hardware Interface Development
- Software Development
- Human Factors
- Payload Safety Review
- Operations Integration
- Testing (Fig. 6.4)
  - Leak tests
  - Electromagnetic capability
  - Interface verification testing
  - Flight load verification
  - Mission sequence test
  - Integrated compatibility test
  - Subsystems test

- Hardware Delivery
- Bench Review
- Certification of Flight Readiness
- Payload launch site support
- Launch
- Suborbital Operations
- Landing and Return of Payload
- Post-flight Tasks

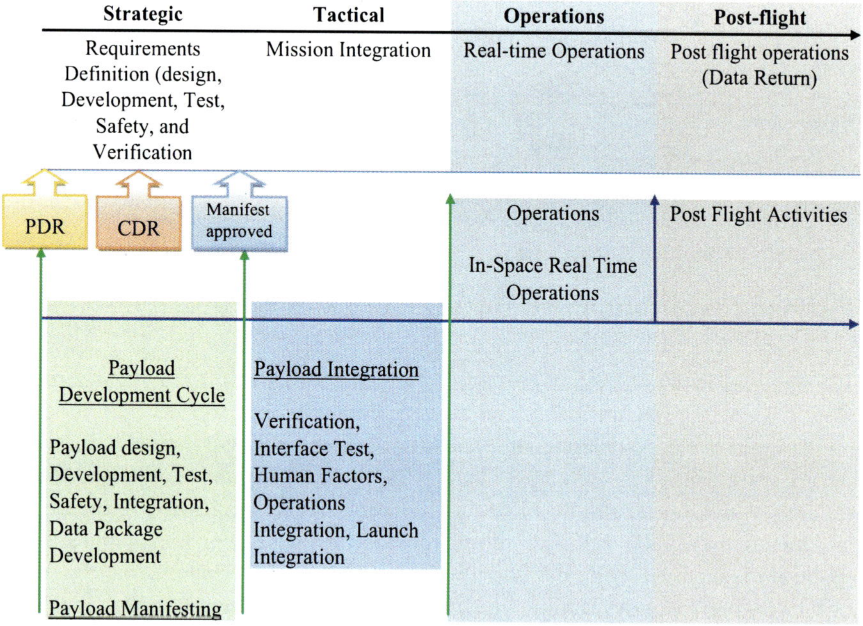

Fig. 6.3 Payload integration process. Image by the author

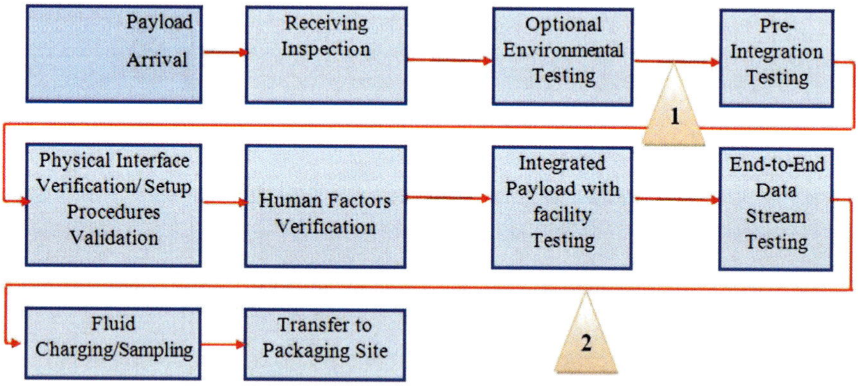

Fig. 6.4 Payload testing process flow. Key: (*1*) Test readiness review. (*2*) Post test review. Adapted from: Payload Developers and Principal Investigators Payload Planning, Integration and Operations Primer. International Space Station Program. ISS National Laboratory. Office June 2010. Image by the author

Spaceport Support Requirements

Needless to say, spaceport support requirements for payloads vary depending on the type of payload. Perhaps the most demanding processing requirements are those for biological sciences experiments since these demand time-sensitive activities such as delivery of samples and also require consumable supplies, chemicals, biohazard waste management, and temperature-monitoring.

Factors to be Considered for a Payload/Science Experiment

Cabin Pressure under Normal Operation
Cabin Pressure under Abort Operations
Cabin Rate of Pressure Change under normal operation
Repressurization/Depressurization rate during nominal and off-nominal operations
Particulate Cabin Air Concentration
Carbon dioxide and oxygen concentration
Temperature during nominal operations.
Temperature ranges during all mission phases.

Payload Characteristics

Payloads can be a human wearing a spacesuit, a couple of middeck lockers stacked on top of one another, or perhaps a science experiment comprising animals. In the latter case some form of animal enclosure module (AEM, sidebar) is required.

Animal Enclosure Module
Those planning on sending animals into space need to ensure the AEM is a self-contained habitat capable of providing its occupants with living space, food, water, ventilation and lighting. And, since the payload may be sitting on the ramp waiting to be loaded, the AEM must feature an internal waste management system and a means to prevent any waste from escaping into the confines of the cabin. Another important feature is temperature: to ensure test subjects are comfortable, fan blowers must be installed. To ensure the air quality is maintained within certain limits, a high efficiency particulate air (HEPA) filter is a feature of the AEM to prevent any microbiological contaminants escaping into the cabin. Water is provided via drinking valves linked to flexible bladders and food is delivered in the form of a sterilized laboratory formula.

Structural Interfaces

Regardless of what type of payload is flown it must survive not only the launch and re-entry loads but also the vibration that occurs during a spaceflight. Also, the vehicle may manoeuvre during the microgravity phase of the flight through the use of the reaction control system. While these accelerations are much smaller than those of launch and re-entry, they still represent a design condition for the payload. And, if the flight goes pear-shaped and the vehicle has to execute an emergency landing, the payload must be designed to survive emergency landing loads. But it is no good designing the payload to survive the minimum acceleration or landing loads: to ensure the payload is accepted at the safety review it will need to have an ultimate factor of safety of at least 2.0 and perhaps as high as 4.0. That means a payload must survive intact—no fracturing or structural failure—at four times the vehicle's acceleration and landing loads. This is usually achieved using distortion-tolerant foam padding and fixing net retention devices/interfaces.

Environmental Conditions

This is another check in the safety review and yet another spaceport facilities requirement. Cleanliness levels for payloads waiting to be loaded on board a spacecraft can be found in NASA's Contamination Control Requirements Manual, published in February 2012 by the agency's Safety and Mission Assurance Directorate. Once the payload is surgically clean, the payload integrator must demonstrate it can safely contain any by-product of an experiment, liquid or a solid and also that there are no toxic materials that might be discharged into the cabin.

Electrical Power Interfaces

Chances are the payload will need electrical power, so baseline energy allocation must be determined to minimize power requirements for the simple reason that payloads with high power requirements will reduce the chances of being manifested. To ensure continuous power is fed to the payload throughout the flight an overload protection circuit (provided by a 10 amp circuit breaker) may be fitted. Another item on the electrical power checklist is ensuring the electrical wiring insulation is rated to vehicle's requirements (when the vehicle is on ground power, the payload may experience voltage transients and power ripples).

Electromagnetic Compatibility

If the payload produces an electromagnetic environment shielding will be required to make sure the payload doesn't interfere with the vehicle's operation. Equally, the vehicle will produce an interference environment with radiated electrical fields and it will be important that this radiated interference doesn't affect the payload.

Payload Services and Facilities

For a spaceport to be successful in payload processing it will need to include a wide range of services and facilities that cater not just to satellite payloads, but also passengers (Table 6.1), since these also happen to be classified as payloads in the commercial spaceflight world. But since passengers require very specific facilities, we will deal with these separately in the following chapter. In this chapter we will continue to focus on those facilities required to support launching satellites and science payloads.

To begin with, spaceport payload processing facilities may be fairly basic, but as business picks up some services may become increasingly specialized. For example, spaceports may feature dedicated processing facilities for animal care or advanced materials or Cubesats. In the latter example, a customer processing a cube satellite will need the use of a clean room (sidebar), a thermal vacuum chamber, a vibration table, an acoustic chamber and an electronic bench, all of which must be housed locally at the spaceport. In contrast, a customer who needs to process a more substantial payload and/or science payload will need the following more substantial spaceport services:

- Clean room (sidebar) and protective space for science payloads
- Payload container system
- Ground support equipment specific to the spacecraft
- Hazardous waste management facilities
- Work benches, storage area and work area for flight preparation and overhaul.
- Engineering maintenance
- Tools and support equipment
- Breathing oxygen (for animal enclosures for example)

Clean Room Classification

A clean room is an extremely clean environment that is achieved by controlling filtration, ventilation, humidity, temperature, air pressure and ionization to very high standards. The quality of a clean room is determined primarily by the concentration of particles in the air. The measurement of particles is the micron, which is one millionth of a meter. Classifying clean rooms in the U.S. is done by applying Federal Standard 209–1992. Using this classification, a Class 8 (Federal Standard Class 100,000) clean room is one in which there are no more than 3,520,000 particles of 0.5 μm per cubic meter of air. In such a clean room, masks and gloves are not required. Higher up the classification is the Class 6 (Federal Standard Class 1000) which permits no more than 35,200 particles equal to and larger than 0.5 μm per cubic meter. In this level of clean room gloves and facemasks are used. At the far end of the classification is the Class 3 (Federal Standard Class 1) clean room which has no more than 35 particles equal to or greater than 0.5 μm. This type of clean room is so clean that humans don't work in the environment, no matter how sterilized they are.

Table 6.1 Space tourist vs. payload processing

Space tourist processing		Payload processing	
Terminal facility activities: Begins with reception before progressing to training, launch and post-mission celebration Processes: exclusivity maintained to varying degrees for space tourists and their family/friends		Terminal facility activities Flight preparation of science payloads Training Ground support activities Payload testing and processing/ bench testing Use of clean room	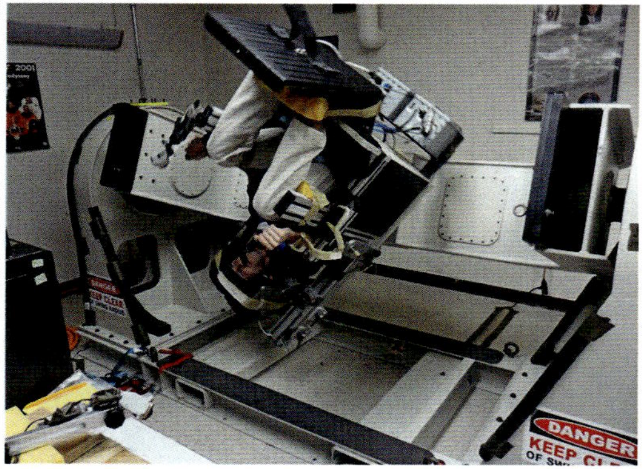

Hangar area activities are driven by turnaround schedule that may exceed one flight per day: – Installation of rocket motors – Fuelling Thermal protection system overhaul Oxygen servicing Nose skid cone replacement Data download Cleaning of cabin Protective covers on windows		Hangar area payload ingress/egress – Installation of rocket motors – Fuelling Thermal protection system overhaul Oxygen servicing Nose skid cone replacement Data download Cleaning of cabin Protective covers on windows	

Images courtesy of Caribbean Spaceport, NASA, Astronaut for Hire, and Reaction Engines

Needless to say, the business of payload processing is a complex one, and spaceports that support payload processing *and* passenger processing will need to consider the impact of one upon the other. For example, the activities within the spaceport building that support the space tourist will be somewhat different than the facilities that support the payload specialist: space tourists will be more interested in food, entertainment and post-mission celebration, while the payload specialist will need protective spaces for flight preparation, payload staging areas and ground support equipment. The bottom line is that the concept of payload operations may conflict with the flight profile for space tourists and these differing flight processes may impact ground, pre-flight and postflight operations and turnarounds

Chapter 7
Passenger Training and Training Facilities

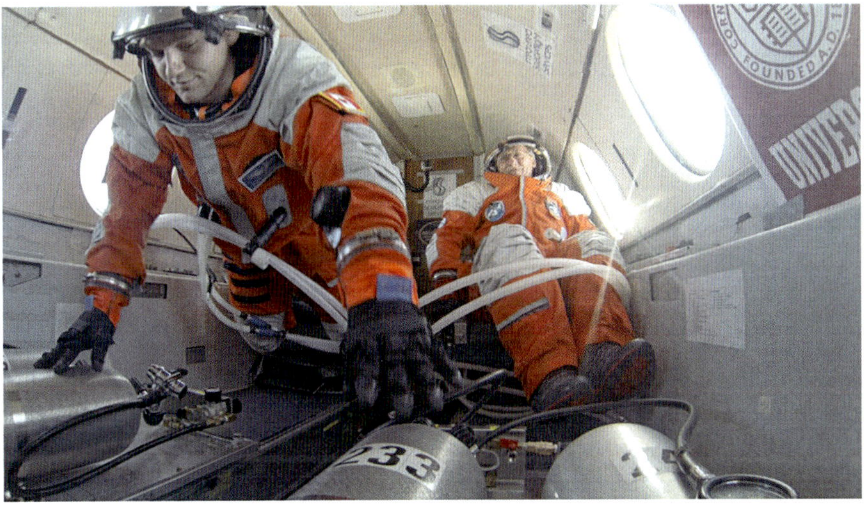

Fig. 7.1 Astronaut candidates/spaceflight participants experience zero-G during a parabolic flight campaign in Canada. Image courtesy of Project PoSSUM

Ideally, spaceports supporting spaceflight participant flights should provide the full spectrum of facilities to support the training activities for those making their journey into space (Fig. 7.1). But what will these activities entail? While taking care of the initial reception of passengers, providing food service and maintaining the exclusivity the wealthy expect will be easy to achieve, the issue of training may be more challenging. That's because different categories of spaceflight participants will demand different types of training. For example, the payload specialist will require an emphasis on procedures and mission rehearsal, whereas the 'space tourist' may want to spend more time being taught how to take Facebook-worthy photos (Fig. 7.2). Further over the

E. Seedhouse, *Spaceports Around the World, A Global Growth Industry*,
SpringerBriefs in Space Development, DOI 10.1007/978-3-319-46846-4_7

horizon there is the difference between suborbital and orbital training. There is also the question of how much training spaceflight participants require? Most suborbital operators such as Virgin Galactic and Blue Origin have settled on a training duration of 2–3 days. Orbital training duration has yet to be defined: the recent crop of visitors to the International Space Station (ISS) trained for 6 months, but when orbital visits become more of a routine this training duration will almost certainly be reduced. In 2016, regular suborbital manned flights are closer to the horizon than orbital, so this chapter focuses on the type of training and training facilities required to cater to this business segment. We'll begin with a generic training schedule as outlined below.

Fig. 7.2 Earth from space. Image courtesy of NASA

Generic Training Schedule for Suborbital Spaceflight Participants

Day One AM

Classroom:

Regions of the Atmosphere
Altitude Physiology and the Hypobaric Chamber
Unusual attitude flight profiles and >Mach 1.0 flight

Day One PM

Chamber:

Rapid and Slow Decompression

Classroom:

Acceleration Physiology and the Anti-G Straining Maneuver (AGSM)
Spacecraft safety and emergency egress
Vehicle indoctrination: safety systems and mission architecture

Day Two AM
Classroom:
AGSM Review, theory and practical.
Centrifuge:
Gradual onset runs (GOR) # 1 and 2 (familiarization to 6 G)
Rapid onset runs (ROR) # 1–3 (ROR4 for 15, ROR5 for 15 and ROR6 for 15)
Debrief/review of G-videos

Day Two PM
Classroom:
Vehicle life support systems
Spacesuit Indoctrination
Spacesuit Donning and Doffing (practical)
Spacesuit Ingress and Egress (simulator)
Chamber:
Armstrong Line chamber run to 80,000 ft wearing spacesuit.

Day Three AM
Classroom:
Flight briefing
Unusual attitude and high-G flight in Extra 300. 45 min

Altitude Physiology and HAI

One of the academic sections delivered to prospective spaceflight participants is altitude physiology and aligned with this is high altitude indoctrination (HAI) training. HAI is conducted in a hypobaric chamber (Fig. 7.3) capable of 'flying' to 100,000 ft altitude and supporting rapid decompression rates from sea-level to 100,000 ft in less than 5 s. These chambers are the perfect environment for all manner of spaceflight participant training, including simulating flight profiles, practicing emergency procedures, evaluating suit performance in low pressure conditions and building confidence while wearing a pressure suit during rapid or slow decompression (RD/SD).

When it is time for their chamber flight spaceflight participants don helmets and mask, hook up their masks and check communications and oxygen systems. The inside observers (IO's) complete a physical check of everyone's oxygen connection to make sure the connections are secure. This is followed by a communication check beginning with the students and ending with the chamber crew. With final checks complete the hatch is closed and the flight director (FD) gives a thumbs up, indicating the flight is about to begin. Once the FD receives a thumbs-up from each student the chamber begins its ascent to 5000 ft at a rate of 5000 ft per minute. At 5000 ft the chamber levels off and the IO's asks each student to complete an ear and sinus check followed by a confirmatory thumbs up. If everyone has clear sinuses the chamber returns to sea level and the IO's once again ask for a thumbs-up to make sure everyone is ready for the hypoxia demonstration flight to 25,000 ft. After levelling off at 25,000 ft, students are divided into two groups. The first group drop

Fig. 7.3 Spaceflight training inside a hypobaric chamber. Image courtesy of ESA

Table 7.1 Hypoxia symptoms

Stages	Indifferent 90–98 % oxygen saturation	Compensatory 80–89 % oxygen saturation	Disturbance 70–79 % oxygen saturation	Critical 60–69 % oxygen saturation
Altitude	0–10	10–15	15–20	20–25
Symptoms	Decrease in night vision	Drowsiness Poor judgment Impaired coordination Impaired efficiency	Impaired handwriting Impaired speech Decreased coordination Impaired vision Impaired cognitive function Impaired judgement	Circulatory failure CNS Failure Convulsions Cardiovascular collapse Death

their oxygen masks after being told to put on their oxygen masks as soon as they experience one clear cut symptom of hypoxia. While off oxygen the students perform simple tasks such drawing cats and dogs, subtraction and multiplication, and answering general knowledge questions. These tasks are intended to demonstrate to the students just how insidious hypoxia can be. To help the inside observers monitor symptoms, the group experiencing hypoxia wear pulse oximeters that display oxygen saturation. Once everyone in the first team has experienced one clear cut symptom of hypoxia and is hooked up to their masks it is the turn of the second group to drop their masks and the exercise is repeated. After the flight the students are quizzed about their hypoxia symptoms (Table 7.1).

The day after their hypoxia demonstration the students enter the chamber for a second flight to 43,000 ft. The purpose of this flight is to demonstrate positive pressure breathing and also to provide another opportunity for students to practice clearing their ears during descent. At 10,000 ft students are told to drop their masks, although the IO's keep their masks on until the chamber reaches ground level. Two chamber runs down, two to go. The third chamber run is the RD flight which is intended to simulate an immediate loss of pressure caused by a puncture in the skin of the vehicle and/or spacesuit.

Before describing what happens during the RD it is helpful to understand the layout of the chamber. The chamber has two sections, one of which is the main chamber and the other secondary chamber or lock. The chambers can operate independently thanks to a hatch and valve separating the two, which means the main chamber can operate at a different pressure than the lock. During the RD run the main chamber is sealed and 'flown' to 40,000 ft. As the main chamber is being 'flown', the IO and students enter the lock, complete their hook-up and checks and 'fly' to 8000 ft. As they ascend the IO reminds the students what they can expect during the RD, and once the lock reaches 8000 ft the chamber engineer prepares to open the valve separating the two chambers. Once the IO has briefed the students to expect a RD, all the students can do is wait. But not for long. After a few seconds the chamber engineer pushes the button opening the valve. A moment later there is a loud bang followed by fogging and a rush of wind as the air in the lock is rapidly evacuated. After they have recovered from the shock, the students busy themselves checking their connections as instructed before giving the IO the required thumbs up. With the RD over all that remains is the final and fourth flight to 80,000 ft (sidebar), which takes place after the spacesuit indoctrination.

Spacesuit Indoctrination

> We have always been clear that a shirt-sleeve environment was part of the baseline design. However, safety remains the priority, and should any new factors emerge that mean we should change that or any other element, then of course we will do so.
>
> Virgin's Stephen Attenborough in a 2013 magazine article

> What we do know is that even if Siebold did not experience ebullism, future space tourists—in the event of a cabin depressurization or spacecraft breakup—could. That's because the space industry has defined the 'outer edges of space' as 62 miles, or nearly 100 km, well past the Armstrong limit (the point at which water boils at 98.6° Fahrenheit, the temperature of the human body). The possibility of ebullism (and other pressure-related elements) drives home the need for all passengers to don pressure spacesuits and oxygen masks, not T-shirts and shorts like some idealized visions of consumer space travel.
>
> Michelle La Vone, Space Safety Magazine, December 2014

In 2008 Nassim Nicholas Taleb published 'The Black Swan: The Impact of the Highly Improbable' and so the 'black swan' event was born. A black swan event is one that is rare and difficult to predict, such as the 1987 stock market crash, the September 11 attacks, or the SpaceShipTwo accident in 2014. Ever since the X-Prize-winning flight of SpaceShipOne in October 2004, Virgin Galactic pilots have worn one-piece flight suits because the thinking was the pressurized cabin would be sufficient protection against the elements. Should they have worn pressure suits? Virgin Galactic's position on pressure suits seems contradictory given the history of manned suborbital spaceflight because NASA and the USAF never considered flying their vehicles without their pilots wearing a pressure suit. So why did Virgin Galactic think it would be safe? One factor that may have swayed Virgin Galactic was the work attire of astronauts on board the ISS because those working on the orbiting outpost wear nothing more than shirtsleeves and pants, an image that reinforces the perception that flying to space is safe. Of course, it is anything but, because government-employed astronauts are among the most highly trained humans on and off the planet, and there are myriad emergency protocols in place to protect crewmembers in the event of an off-nominal event. Virgin Galactic wants to make spaceflight fun, but that will prove difficult if passengers are tethered to oxygen hoses and constrained by pressure suits. But in light of the SpaceShipTwo accident the pressure suit issue is one operators will likely adopt, so spacesuit testing and training (Fig. 7.4) facilities will be required at spaceports.

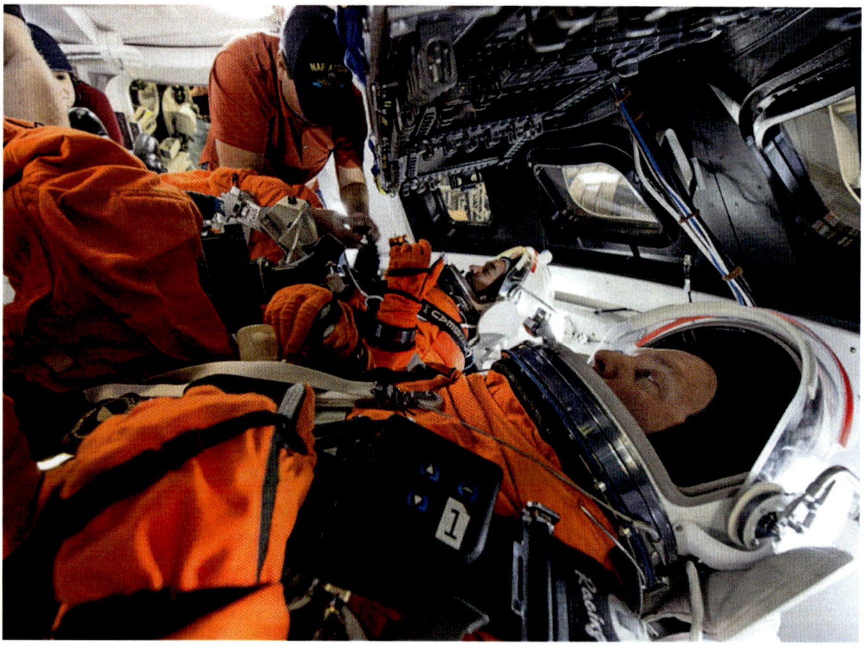

Fig. 7.4 Spacesuit testing in a mock-up capsule, a capability that will be required by spaceports supporting passenger training. Image courtesy of NASA

With spacesuit training checked off, trainees will perform their final chamber flight. One of the most important objectives of all this chamber training is to expose spaceflight participants to simulated altitude so they learn about their limitations and dangers of working in what is a very, *very* unforgiving environment. It also provides an ideal opportunity for spaceflight participants to familiarize themselves with the spacesuit they will be wearing on their flight. After donning the spacesuit the spaceflight participants ascend to 80,000 ft while listening to instructions from the FD. As the chamber ascends to altitude the spaceflight participant is told how the suit should feel as it inflates. As the chamber passes 63,000 ft the spaceflight participant notices the suede patches on the suit beginning to smoke. At 80,000 ft the spaceflight participant performs simple tasks such as pulling a pen from a pocket, checking D rings, performing a simple press to test, and going through the emergency checklist. Once the spaceflight participant has completed the tasks they are left at altitude for a few minutes so they can fully appreciate the potentially lethal situation they are in and also to build confidence in the suit. With the familiarization to 80,000 ft over, the chamber descends to 25,000 ft and levels off in preparation for the final chamber test: with the punch of a button on the chamber console, a loud bang and a rush of air, the two chambers equalize at around 70,000 ft. After their second RD in as many days, the spaceflight participants descend at 5000 ft per minute to ground level, they doff their suits, post-flight records are written, and a briefing is conducted.

Acceleration Physiology

In the academic module spaceflight participants are taught about the different types of G force. Gx is the force that acts from chest to back and is experienced during take-off, Gy is the lateral force that is familiar to aerobatic pilots when they perform aileron turns, and Gz is the force that acts through the vertical axis of the body, from head to foot or from foot to head: if Gz is experienced from head to foot it is termed positive Gz (+Gz) and if acceleration is transmitted from foot to head it is termed negative Gz (−Gz). These G forces exert a significant strain on the body, particularly the cardiovascular system, which must keep blood flowing to the brain (sidebar). While the cardiovascular system responds quickly to increased acceleration by increasing heart rate, there is a point at which the physiological responses cannot keep pace with the Gs. When that happens, the cardiovascular system cannot pump sufficient blood to the brain and pilot performance is degraded, sometimes with fatal consequences. One of the first signs that things are going pear-shaped is loss of vision (LoV) because the eyes are particularly sensitive to low blood flow. Fortunately, the symptoms of high G exposure are fairly predictable, and pilots are trained (sidebar) to recognize these, and under normal flight conditions there is a low risk of either the pilot or his passenger experiencing a problem, but if things go squirrely and those Gs pile on rapidly, acceleration training will be invaluable.

AGSM

The AGSM is a technique taught to all fighter pilots to increase their tolerance to the dreaded G. A perfectly executed AGSM will increase your tolerance by two to three G, so spaceflight participants will need to become proficient. The technique is all about timing and requires a deep inhalation of air while simultaneously tensing the big muscles in the legs, stomach and buttocks. Following a count of three the pilot exhales rapidly, inhales again and repeats the exercise.

The few centrifuges that exist vary widely in their capabilities but perhaps the gold standard in the centrifuge world is the Phoenix 4000 at NASTAR. Located in Southampton, Pennsylvania, this luxury 'fuge' (Fig. 7.5) is extremely versatile, which is one reason it has been used by Virgin Galactic to train its spaceflight participants. In standard aircrew training, pilots are required to complete a series of gradual onset runs (GOR), rapid onset runs (ROR) and high sustained G (HSG) runs: a GOR is defined as an onset rate of 0.1 G per second, a ROR is defined as an onset rate of at least 3 G per second, and a HSG run is run in which a pilot is subjected to 7 G for 15 s wearing a G-suit or 5 G for 15 s without G protection. For spaceflight participants the acceleration forces will not be quite so substantial, but centrifuge training will still be required to help this new class of space passenger better understand the physiological stresses of increased G, and to increase their confidence in their ability to tolerate high Gs.

Fig. 7.5 NASTAR's centrifuge. Image courtesy of NASTAR/ETC

As to the question of whether spaceports should co-locate centrifuges, the cost may be prohibitive, since a centrifuge in 2016 dollars will set a spaceport back $22 to $28 million.

Space Motion Sickness

Nausea, vertigo, headaches, vomiting, and general discombobulation. These are all symptoms (Table 7.2) of space adaptation syndrome or space motion sickness (SMS). Typically, more than half of first-time astronauts suffer from SMS, and it will be helpful for spaceflight participants to understand the problem before they fly.

Table 7.2 Motion sickness symptoms and criteria for grading motion sickness severity [1]

Category	Pathognomonic (16 points)	Major (8 points)	Minor (4 points)	Minimal (2 points)	AQS (1 point)
Nausea syndrome	Vomiting or retching	Nausea II, III	Nausea I	Epigastric discomfort	Epigastric awareness
Skin		Pallor III	Pallor II	Pallor I	Flushing
Cold sweating		III	II	I	
Increased salivation		III	II	I	
Drowsiness		III	II	I	
Pain					Headache
Central nervous system					Dizziness Eyes closed>II Eyes open III

One serious consequence of a spaceflight participant's sensory system being out of sorts is responding to an emergency since, with the crewmember's perceptual-motor system compromised, the spaceflight participant will find it difficult to perform tracking and fine manipulation tasks. Given the negative operational consequences of SMS it isn't surprising that space agencies have spent a lot of time and resources developing countermeasures. The Russians had some success with pre-flight stimulation of the vestibular system and similar pre-flight behaviour modification techniques, which spurred an interest on the application of pre-flight adaptation training techniques such as virtual reality. The theory behind this type of training (sidebar) is that devices such as virtual reality can simulate the sensory realignment that occur in microgravity that cause SMS: by repeatedly exposing astronauts to unusual environments it should be possible for crewmembers to encode and adapt to these challenging stimuli.

Preflight Virtual Reality Training

Astronauts engaged in this type of training typically perform standard mission tasks, such as navigation or switch activation, in multiple orientations in a virtual environment in the hope they will retain the skills acquired when it comes to the real thing. Over the years the training has proven to be quite effective, with the number of crewmembers suffering from SMS symptoms reduced by half. Another similar training method is autogenic feedback training (AFT) which employs psycho-physiological countermeasures to condition crewmembers to voluntarily control their physiological responses. The training, which takes place over several days, involves a lot of repetition and practice, but at the end of the conditioning, most people are able to exert a greater control over their physiological responses to motion stimulation.

Perception

In addition to the problems of SMS, spaceflight participants must contend with the effects on their perception, which is disrupted as a result of illusory changes during the flight. For example, there may be many who will find the *inversion illusion* troubling, which means it will be important to focus on strong orientation cues and secure restraints. One strategy for dealing with this is adaptation and pre-adaptation. For spaceflight participants who may be flying just once, the best way of adapting is by using parabolic flight. There are different rates of adaptation, but generally it takes up to 3 days or 3–4 min at zero-G per day for sensorimotor changes to develop. This means spaceflight participants will need to fly one parabolic flight (see below) sequence per day for 3 days before their flight. An alternative is using short-radius centrifugation and/or unusual attitude training in a high performance jet fighter.

Parabolic Flight

Although a parabolic (Fig. 7.6) flight only provides 20–24 s snapshots of what being weightless feels like, it is an invaluable adaptation training tool for suborbital spaceflight. Once the aircraft reaches 350 knots indicated airspeed (Mach 0.83) and an altitude of 7300 m, a gradual climb is initiated at full thrust thereby generating vertical speed without sacrificing airspeed. During the gradual climb the G-meter reads 1.5 G, but this reading increases to 1.8 G as the pitch angle increases to 45° (the 'pull-up'). At 225 knots indicated airspeed, with the aircraft closing in on an altitude of 10,000 m, the pilot begins the parabola by pushing forward on the control yoke. This lowers the angle of attack of the wings, which in turn reduces wing lift. As the power is simultaneously reduced, airspeed falls as the aircraft reaches the top of the parabola, which it reaches at 10,000 m (Mach 0.43 or 140 IAS/245 TAS—this speed isn't much faster than the stall speed). This is when passengers start floating around the cabin.

a

b

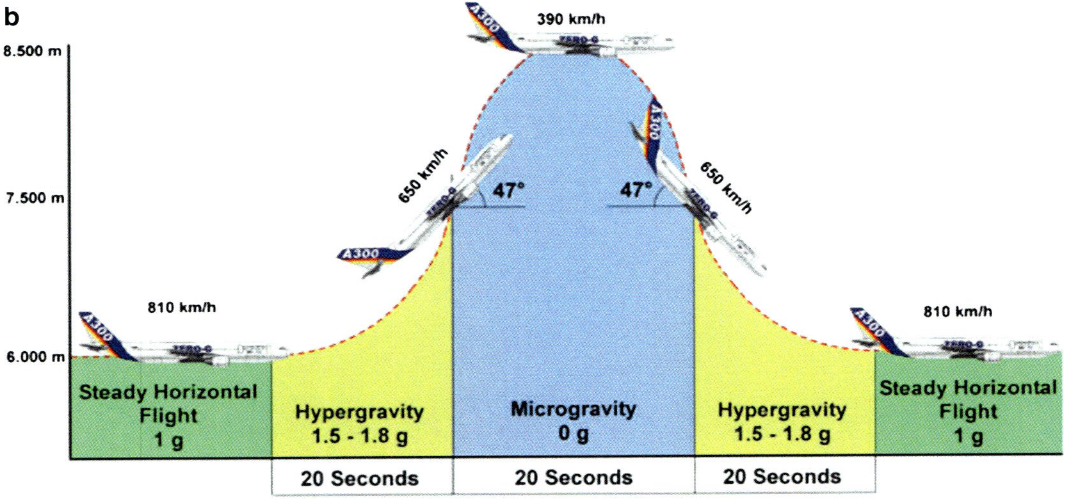

Fig. 7.6 Parabolic flight training allows spaceflight participants to experience 22–24 s of microgravity per parabola. (**a**) ESA/CNES parabolic flight aircraft during a parabolic flight campaign. (**b**) A standard parabolic flight profile. Images courtesy of ESA

Reference

1. Graybiel, A., Wood, C.D., Miller, E.F., Cramer, D.B.: Diagnostic criteria for grading the severity of acute motion sickness. Aerosp. Med. **39**, 453–455 (1968)

Chapter 8
Point-to-Point Transportation

Fig. 8.1 A conceptual reusable suborbital passenger vehicle. Image courtesy of EADS Astrium

In the not-so-distant future, if you believe the articles in Popular Mechanics or Wired, the more affluent sections of society will have the option of rocketing between London and Sydney or between New York and Beijing in just 2 h (Fig. 8.1). Or less. It's called suborbital point-to-point (SPTP) [1] and, if those articles or to be believed, it will radically alter intercontinental commuting. At least for a privileged few. With the advent of suborbital reusable launch vehicles (sRLV) comes the potential for PTP transportation which has long been heralded as a revolutionary

E. Seedhouse, *Spaceports Around the World, A Global Growth Industry*,
SpringerBriefs in Space Development, DOI 10.1007/978-3-319-46846-4_8

mode of air transport. One study published by the International Space University (ISU) in 2008 calculated the transatlantic flight time for a sRLV between New York and London could be as quick as 75 min. While such rapid global transport of passengers makes PTP an attractive space technology proposition, its success is dependent on a number of factors. First, PTP transport needs to be integrated into not only the National Airspace System (NAS) but also into International Airspace. Second, there are myriad institutional, technical and operational issues that must be resolved before supersonic and hypersonic PTP systems (Fig. 8.2) can be flown between major cities. And third, dedicated spaceports may need to be co-located [2] with existing airports.

Fig. 8.2 One hypersonic vehicle being developed is the Skylon by Reaction Engines. It may be operational in the late 2020s if all goes well. Image courtesy of Reaction Engines

Spaceport Location

It is this third item that is the key to realizing a potential PTP industry because for these flights to be commercially-viable they must be flown between densely-populated international hubs. This is because the whole point of PTP is to significantly reduce the time it takes for passengers to reach long distance destinations. But if a spaceport, such as Spaceport America, is situated at a remote location, the transportation time to and from the spaceport must be factored into the equation. In the case of Spaceport America, which is a 2 h drive from El Paso or a 6 h plus drive from Phoenix, the savings in journey time are offset by the trouble getting to the spaceport in the first place. The PTP spaceport must also provide facilities for all RLV types and also all the passenger services normally provided by an airport. The aforementioned ISU report identified likely population-dense locations (Table 8.1) that might support PTP passenger routes, while a study by the FastForward Study Group grouped potential demand by tiers as depicted in Table 8.2.

Table 8.1 ISU PTP routes

Rank	Route	Rank	Route
1	Los Angeles–New York	11	Beijing–New York
2	New York–London	12	Hong Kong–New York
3	Tokyo–New York	13	Chicago–Paris
4	Tokyo–London	14	Beijing–London
5	Chicago–Tokyo	15	Hong Kong–London
6	Chicago–London	16	Frankfurt–New York
7	Los Angeles–Tokyo	17	Singapore–New York
8	Paris–New York	18	Frankfurt–Tokyo
9	Paris–Tokyo	19	Los Angeles–Paris
10	Los Angeles–London	20	Singapore–London

Table 8.2 Fastforward PTP routes

Feasible city pair route counts by origin			
City	Tier 1 Routes	Tier 2 Routes	Tier 3 Routes
Los Angeles	2	2	2
New York	2	2	4
London	2	2	4
Cologne	2	4	5
Shanghai	1	4	4
Hong Kong	0	3	3
Tokyo	0	4	4
Mumbai	0	5	6
Dubai	0	5	6
Sydney	0	3	3
Buenos Aires	0	0	2
Sao Paulo	0	0	2
Johannesburg	0	0	5
Total	10	36	50

PTP Challenges[1]

Thermal Protection System

When we are talking about traveling intercontinental distances at very high speeds it is worth considering the velocity—or delta-V—required. Say you are planning a 10,000 km trip from the United States to Europe. This journey would require a delta-V of more than 7300 m per second and the amount of fuel required to achieve that would be significant. And, since the vehicle would be travelling so fast, it would require a very robust thermal protection system (TPS) to survive re-entry. In fact, once you start operating vehicles at hypersonic speeds, the sRLV takes on more and more of the characteristics more commonly associated with an orbital RLV.

Super-Density Operations

A PTP vehicle will have significantly different performance characteristics than a commercial jet. For example, a PTP vehicle will have a steep glideslope, may re-enter the Earth's atmosphere as a glider and will therefore have very limited maneuverability. This presents a conundrum because how will such a vehicle participate in carefully-planned air traffic flows [3], even with all the latest NextGen advanced modeling, navigation and intercommunications technologies?

Passenger Accommodation

Aviation administrations are tasked with keeping passengers safe [4] while flying commercial. To that end they are tasked with certifying passenger restraints, emergency procedures and training cabin crew. For PTP operations all these procedures will need to be revised to deal with weightlessness, operating in confined spaces, and launch and re-entry G forces.

[1] More detailed information about PTP vehicles can be found in the following publication: Point-to-Point Commercial Space Transportation in National Aviation System Final Report, March 10, 2010. U.S. Department of Transportation Research and Innovative Technology Administration Volpe National Transportation Systems Center. Prepared for: Office of Commercial Space Transportation U.S. DOT Federal Aviation Administration 800 Independence Avenue, SW Washington, DC 20591. Prepared by System Business Design & Deployment Division Volpe National Transportation Systems Center, U.S. DOT 55 Broadway, Kendall Square Cambridge, MA 02142.

High Altitude Winds

System performance of PTP vehicles with respect to accuracy, precision and reliability when exposed to wind shear/high altitude winds [5] is an unknown quantity when it comes to PTP vehicle performance.

Environmental Impact

Rockets are noisy, but noise [6] isn't the only environmental impact spaceports will need to deal with. There are also the not inconsiderable issues of greenhouse gas emissions, carbon footprint impact, sonic shock waves, vibration, and the effects of the chemical components of spacecraft emissions.

PTP Market

There have been no credible market studies that specifically address the use of spaceports in support of PTP. Not only that, but the optimum design for a spaceport capable of supporting PTP has not been defined, nor has the technical design been determined. None of these issues (Table 8.3) are surprising since the price points are uncertain and it will be a long while before they are clearly defined. But the key to the use of spaceports in support of PTP are the passengers. Who will they be and what will

Table 8.3 Challenges to PTP transportation

1.	Access » Airspace Management and Integration with the National Airspace System (NAS) » Establishing Flight Corridors (noise, safety and environmental compliance)
2.	Coherent Legal & Regulatory Framework For International Operations » Operate under License and Informed Consent » Insurabilityn of Operations » Cooperative Regulations (FAA, CAA) » Legal Operating Frameworks for Domestic and International Spaceports
3.	Common Spaceport Procedures » Fuel Handling and Storage » Customs Issues
4.	Availability of Spaceport Locations » Investment in Spaceport Infratsrtuctre in Multiple Locations
5.	Potential ITAR Issues » Suborbital vehicles cannot be exported since they fall under the International Trade in Armaments Regulations, or ITAR: because of this, exporting spacecraft will be a very difficult process because it would involve policy decisions that the government will find problematic
6.	PTP Business Model Validation » Showing Relative Advantage of PTP & Spaceports against Existing Airports » Defining Time Savings from PTP (travel time from population center to spaceport

the demand be? Will the demand come from tourists primarily or will businessmen make up most of the passenger manifests? And what about cargo? No one knows.

The point is, just because PTP is technologically feasible doesn't necessarily mean it will happen. And no matter how appealing flying from London to Sydney in 3 h instead of 22 h is, the PTP business model is not viable—at least in the short term. One reason is the inconvenience factor. If you happen to be wealthy enough to afford that $30,000 first class intercontinental seat then you are also able to afford to pay for a seat on a business jet. These jets, which can be leased by the hour and can also be shared, turn up when requested and take the passenger to where the passenger wants to go. These jets are not bound by a schedule, there are no lengthy security checks and passengers don't have to drive to remote locations to the hub. But a hypersonic PTP aircraft must fly along pre-timed flight paths under conditions of very high security and is anything but convenient. But a business jet flies on command and is therefore quicker across intercontinental routes than hypersonic PTP aircraft unless that PTP aircraft flies multiple daily flights which will only happen if operating costs are similar to traditional (subsonic) aircraft. We may have some insight into how successful PTP transportation may be when the new breed of supersonic jets makes its debut in the next few years and also how well NextGen evolves to its mid-term capabilities. If air traffic management procedures can seamlessly integrate suborbital flight operations and if the environmental impact can be reduced, and if the passenger accommodation issues can be resolved, and if—the biggest 'if'—the market can be proven, then PTP suborbital may just take off.

References

1. Federal Aviation Administration: Concept of Operations for Commercial Space Transportation in the National Airspace System, Version 2.0. http://www.faa.gov/about/office_org/headquarters_offices/ast/media/CST_CONOPS_v2.pdf (May 2001)
2. Charania, A.C., Kelly, M., Olds, J.R., FastForward Study Group, Space Works Engineering, Inc.: Aiming Towards Viable Business Models of Global Hypersonic Point-to-Point Cargo Transportation. Copyright ©2009. http://www.sei.aero/eng/papers/uploads/archive/IAC-09.E6.1.9.pdf (2009)
3. Murray, D.P., FAA AST: Space and air traffic management of operational space vehicles, AIAA 2008-6890. In: AIAA Atmospheric Flight Mechanics Conference, Honolulu, Hawaii, 18–21 August 2008. http://www.faa.gov/about/office_org/headquarters_offices/ast/reports_studies/media/AIAA-2008-6890-470%5B1%5D.pdf
4. Mineiro, M.C.: Law and regulation governing US commercial spaceports: licensing, liability, and legal challenges. J. Air Law Commer. **73**(4), 759 (2008)
5. Walterscheid, R.L.: High-Altitude Wind Prediction and Measurement Technology Assessment. Report No. DOT-VNTSC-FAA-09-13, The Aerospace Corporation under contract to the Volpe Center, 30 June 2009
6. DOT/FAA 14CFR Parts 36 and 91, Stage 4 Aircraft Noise Standards, RIN 2120-AH99 [Docket No. FAA-2003-16526, Amendment No. 36-26, 91-288], Effective: August 2005

Chapter 9
Spaceports Around the World

Fig. 9.1 Artist's concept of the Caribbean Spaceport. Image courtesy of Caribbean Spaceport

It's October 2004 and SpaceShipOne (SS1) has just rocketed into space on its historic X-Prize winning flight (Fig. 9.1). The ink was barely dry on the $10 million Ansari X-Prize winning cheque before a potentially equally lucrative space race was announced; the competition between spaceports. Kick-starting the contest was Peter Mitchell, director of the New Mexico Office for Space Commercialization. "Today doesn't belong to New Mexico, the day belongs to the gentlemen up here ... that made this dream become a reality," Mitchell told reporters. "Tomorrow, however, we focus on bringing the spoils of this dream to the state of New Mexico." And so the

E. Seedhouse, *Spaceports Around the World, A Global Growth Industry*, SpringerBriefs in Space Development, DOI 10.1007/978-3-319-46846-4_9 race was on. Or so we thought: fast forward 12 years and there have been no revenue suborbital passenger flights. None. There has also been no activity—of the manned kind at least—at Spaceport America, which was built to serve Virgin Galactic and what should have been welcoming a steady stream of passengers by now.

To be fair, the X-Prize took place way back in 2004, before the suborbital passenger business was even a business. But with market studies predicting suborbital space tourism generating more than billion dollars a year in revenues by 2021, the industry's main players figured space passengers would need a place to train and a suitably high-end resort to stay at. Not the sort of facilities you normally find at rocket-launch ranges, which generally have lots of wide-open space but little else. In short, there needed to be a viable tourist destination. And so the spaceport was born, perhaps the most well-known of which is Spaceport America.

Spaceport America

Owner/Operator: New Mexico Spaceport Authority. Located at 901 E. University Ave, Suite 965 L, Las Cruces, New Mexico, 88001 (it's the building just west of Starbucks, and their office is on the second floor). Tel: 575-373-6110
Location: Sierra County, New Mexico
Hub for: Virgin Galactic, UP Aerospace
Elevation, AMSL 1401 m
Coordinate: 32°59′25″N 106°58′11″W
Runways: 16/34 3657 m (concrete)
Nearest Airports:

- Albuquerque International Airport (ABQ) in Albuquerque, New Mexico. Take Interstate 25 South for 146 miles/235 km before taking Exit 79.
- El Paso International Airport (ELP) in El Paso, Texas. Take Interstate 10 West to Interstate 25 North for 126 miles/203 km before taking EXIT 75.

Accommodation: limited.
Visitor Center: open daily from 8:30 am until 4:30 pm.

www.spaceportamerica.com

If you happen to have a ticket to fly on SpaceShipTwo, then Spaceport America is where you will start your journey. On arrival, you will issued magnetic tags that allow you access to the astronaut lounge, a vast atrium filled with natural light from an elliptical wall of windows, offering a vista of the 3-km-long runway. Built from scratch and designed to accommodate a regular passenger service, Spaceport America (Fig. 9.2) was built from nothing in the middle of nowhere 50 km from the nearest town. And it wasn't cheap; with a price tag of almost a quarter of a billion dollars and counting, the spaceport was paid for by the state of New Mexico, whose citizens voted for a sales tax designed to finance its construction. Thanks to the fact that it is the anchor tenant for Sir Richard Branson's Virgin Galactic, the world's first commercial spaceline, the $200+ million project has attracted worldwide attention.

Fig. 9.2 WhiteKnightTwo and SpaceShipTwo on the taxiway at Spaceport America. Image courtesy of Jeff Foust

Designed, built and operated by the New Mexico Spaceport Authority (NMSA), Spaceport America's operational infrastructure includes an airfield, launch pads, a terminal/hangar facility, emergency response capabilities, utilities and roadways. The spaceport will also be capable of accommodating the activities of vertical and horizontal takeoff vehicles, serving as a base for astronaut training, and providing a tourism experience for families and friends of those with a ticket to ride.

Spaceport America's location is remote to say the least. Located west of the U.S. Army White Sands Missile Range in Sierra County, in New Mexico, or about 50 km southeast of Truth or Consequences, Spaceport America is easily accessible as long as you have a car and are prepared to drive the 150+ kilometers from either of the two nearest airports, El Paso International Airport (ELP) or Albuquerque International Sunport (ABQ). While Spaceport America may look as if it dropped out of the future, its facilities are surprisingly limited: for example, it's not the sort of place you can land your Citation jet because it is designated to operate as a prior permission required (PPR) airport, which means there are no services for general or commercial aviation.

But while no rockets have taken off carrying fare-paying passengers, the spaceport has witnessed some activity thanks to residents such as UP Aerospace, which was the first company to launch a commercial payload from the spaceport. Another significant presence is Lockheed Martin, which has been using the spaceport as a base to test new launch and recovery technologies.

Spaceport Sweden

www.spaceportsweden.com

Email: info@spaceportsweden.com
Twitter: @SpaceportSweden
Phone: +46 (0) 980 80 880. Mon-Fri 09:00–17:00 h CET.
Services:

- Arena Arctica. 5000 square meter hangar that can accommodate aircraft as large as a Boeing 747.
- Esrange Space Center. Provides expertise for rocket launches, balloon flights, vehicle testing and satellite control.
- ICEHOTEL. Offers conference facilities and is partnered with Spaceport Sweden.

Deserts not your style? No problem. Fly to Stockholm and drive north to Kiruna, home of the *Spaceport at the Top of the World.* Given its location, the Swedish town of Kiruna may seem an unlikely place to build a spaceport. At 67.86° latitude, it is 150 km above the Arctic Circle and almost 900 km north of Stockholm. In addition to its extreme grid reference, Kiruna has a number of other disadvantages; it is home to the world's largest underground iron ore mine, a vast expanse of forests, no sunlight for weeks at a time, and temperatures that are ideal for polar bears, but not so good for tourists (the average July high is 7 °C). These shortcomings didn't deter the Swedish government though. In 2007, the government announced an "agreement of understanding" with Virgin Galactic to make Kiruna the company's first launch site outside the United States.

While Spaceport Sweden will be new to many tourists, Kiruna is not completely undiscovered. The town has been home to an array of aerospace activities since the Swedish government established a space research center there in 1964. The center—Esrange—includes a 5600 km^2 range for launching sounding rockets (Fig. 9.3). But it is the prospect of people flying into space that has Kiruna's attention. With typical Virgin[1] panache, the suborbital Kiruna flights are promoted as an Arctic adventure complete with a hotel stay [the IceHotel—www.icehotel.com—a hotel with rooms built out of snow and ice (*snice*) where guests pay up to $600 to spend a night bundled in sleeping bags on reindeer skins in sub-zero temperatures] and snowmobile rides through the wilderness.

[1] One hurdle Virgin Galactic faces is the United States Munitions List (USML) which has placed man-rated sub-orbital spacecraft to its list! This is a problem because any item on the USML requires an export license from the US State Department. Even worse, putting suborbital spacecraft on the USML places them under the restrictive umbrella of the International Traffic in Arms Regulations (ITAR). Why is this being done? The 1990s restrictions were intended to block the flow of space technologies to nations such as China and to maintain US space competitiveness. The upshot of this was that the restrictions harmed rather than strengthened the US commercial satellite industry; US satellite makers were denied access to foreign markets and lower-cost launchers for their products.

Fig. 9.3 Kiruna Spaceport. Image courtesy of ESA

Building a hotel out of snice and charging $600 a night may have struck many to be too outrageous a business plan to succeed, but it did, which is probably why Kiruna is so supportive of Virgin Galactic's plans: don't forget that those rich enough to splash out $250,000 on a ticket will most likely want their nearest and dearest along to share the experience, which adds up to lots of hotel rooms. And when it comes to launching rockets, Spaceport Sweden has the infrastructure in place: Esrange stages 5–10 launches per year and it's been in the business of suborbital flights since 1966, albeit unmanned (the range's MAXUS and MASER rockets offer up to 13 min of microgravity at the peak of their suborbital flights).

Mojave Spaceport

Mojave Air and Space Port
1434 Flightline, Mojave, California 93501
Email: info@mojaveairport.com
Phone: (661) 824 2433

Mojave Airport (Fig. 9.4) is a ground zero for commercial suborbital spaceflight. Housed in a collection of dusty hangars and sheds are Scaled Composites, XCOR Aerospace, Masten, and The Spaceship Company. Located a 2 h drive north of Los Angeles, Mojave Air and Space Port has become one of the most iconic locations in the commercial suborbital industry. Home to 14 space companies, this vast expanse of flat, scrubby desert has witnessed *thousands* of rocket tests, although the only vehicle that has flown into space is SS1 (the first orbital Shuttle flight landed there

Fig. 9.4 Mojave Air and Spaceport. Image courtesy of Kluft

as well). While the Mojave Airport and Spaceport has led the way in the spaceport industry since it was the site of the first privately funded spaceflight with the launch of SS1, the new kid on the block—Spaceport America—has established the template for future commercial spaceports.

Baikonur Spaceport

Location: 45.9 N 63.3 E. 200 km east of the Aral Sea
Total area: 6717 km^2
Facilities:

- Site 110. Used to launch Buran and originally built to support the Soviet lunar program
- Site 112. Used to mate Energia launchers to Buran. The main hangar (MIK) was originally constructed to build the N1 moon rocket.
- Site 251. The Buran landing facility. Features one 4500 m long runway (06/24). Now serves as a commercial cargo airport.
- Site 254. Constructed to service Buran between flights. Adapted to service prelaunch Soyuz and Progress operations.

Notable achievements

- 4 October 1957: Sputnik launched from Baikonur
- 12 April 1961: Yuri Gagarin launched from Baikonur

- 16 June 1963: Flight of Valentina Tereshkova, first woman in space
- 20 November 1998: Zarya, the first piece of the International Space Station (ISS) launched on Proton
- 12 July 2000: Zvezda, the main component of the Russian section of the ISS launched on Proton.

Tours: $800. Includes room and three meals at the Tsentralnaya Hotel.

Impossibly bleak, unfathomably remote and utterly featureless, Baikonur was the perfect place to build a top-secret missile range that eventually became ground zero for the Soviet space program. The spaceport, which was leased to the Russian Federation on 28 March, 1994 for a period of 20 years, features nine launch complexes with 15 launch pads and 11 assembly and test areas in addition to:

- 34 technical complexes for assembly, testing and prelaunch preparation of launch vehicles.
- Three fueling and decontamination stations for fueling spacecraft
- An oxygen and nitrogen plant with a production capacity of 300 t of cryogenic products per day.
- An electric power supply grid, which contains more than 600 transformer substations
- Two airports
- More than 400 km of railroad tracks
- More than 1000 km of motor roads
- 2500 km of communications lines.

The Russian Federation uses Baikonur to fulfil multiple roles as follows:

- Space-based Communications. This role includes providing television and radio broadcasting services, remote sensing, time standards and providing navigation coordinates to ground users.
- Space Programs. Baikonur is Russia's leading spaceport in terms of the number of launches, with between 70 and 80% of all launches taking place from the spaceport. Vehicles have been launched into orbits of 200–40,000 km using light, intermediate and heavy launch vehicles.
- Vehicle Preparation and Implementation. The spaceport features complexes for Proton, Soyuz, Zenit, Tsyklon and Rokot vehicles. These complexes comprise launch and technical facilities for preparing vehicles for launch.
- International Programs. The spaceport is used to support international programs in which Russia participates such as Phobos, Vega, Interkosmos and the ISS: many of the complexes were designed specifically for the ISS.

Plesetsk Spaceport

Location: 63° N 41° E
Area: 1762 km^2
Facilities/infrastructure

- ➢ Six launch complexes with nine launch pads
- ➢ Six assembly-test areas with 37 technical complexes for assembly, testing, and preparation of launch vehicles
- ➢ Two fueling and decontamination stations
- ➢ 152.4 km of power lines
- ➢ Railroads and asphalt-concrete roads
- ➢ Water and heat supply network
- ➢ Pero airport.

As the major launch site on Russian territory, Plesetsk Spaceport plays a key role in providing the country with independent access to space: approximately 60 % of all rocket launches from Russian spaceports are launched from Plesetsk.

Roles:

- ➢ Communications. About 35 % of all the spacecraft supporting Russia's space observation program were launched from Plesetsk, 75 % for communications and television broadcasting, approximately 25 % for navigation, 70 % for remote sensing, 60 % for Russia's weather satellite program, and 70 % for the scientific research program.
- ➢ National Space Programs. These include Kosmos, Molniya, Meteor, Foton, AUOS and Prognoz.
- ➢ International Missions. MAS-1, Gelisat, Tubsat, Astrid, COSPAS-SARSAT, Bion and Magion have been launched from the spaceport (Fig. 9.5).

Mid-Atlantic Regional Spaceport

Fig. 9.5 The Orbital Sciences Corporation Antares rocket is seen on the Mid-Atlantic Regional Spaceport (MARS) Pad-0A, April 16, 2013 in Virginia. Image courtesy of NASA/Bill Ingalls

Operator: Virginia Commercial Spaceflight Authority (VCSFA)
4111 Monarch Way, Suite 303 | Norfolk, VA 23508
Phone: 757-440-4020 | Fax: 757-440-4023
E-mail: info@vaspace.org
Launch azimuths: 90° to 160°
Orbital inclinations: 38° to 60°.
Trajectory Options: Polar and sun-synchronous orbits

Location: Wallops Island, VA
Latitude: 37.8° N
Longitude: 75.5° W
Website: www.vaspace.org

The Virginia Commercial Space Flight Authority (VCSFA) began its lease at Wallops Island in 1997 and gradually added to the facilities (Table 9.1) over the years. Today it features one mid-class and one small-class launch facility together with vehicle and payload processing integration facilities through agreements with NASA. The spaceport, which has a legacy of 16,000 launches over 55 years, is one of only four licensed by the FAA Office of Commercial Space Transportation for orbital launches.

Table 9.1 Mid-Atlantic Region Spaceport launch pads

➢ Pad 0A. Mid-Class Launch Facility with:
• Cryogenic Liquid Fueling Facility
• Deluge System for acoustic suppression
• Launch Mount
• Flame Trench
• Hydraulic System (part of the Transporter Erector Launcher system)
• Environmental Control System for the Payload/Launch Vehicle
• Liquid oxygen Subcooler
Pad 0A was originally built for the Conestoga rocket that made only one flight in 1995. The launch tower was demolished in 2008 and rebuilt for Orbital Sciences Antares
➢ Pad 0B. Small-Class Launch Facility:
• Can accommodate solid fuel Launch Vehicles, including Minuteman and Peace Keeper
• Small liquid and hybrid fueled launch vehicles
• Piling-reinforced launch mount/flame duct
• Piling-reinforced concrete apron
• Moveable Service Structure
• Support Equipment Building
• Launch Equipment Vault
Became operational in 1999. Upgraded in 2004 and is used to launch Minotaur rockets

Spaceports in Development: Houston

Operator: Houston Airport System
Arturo Machuca
General Manager, Ellington Airport
(713) 847-4219
Arturo.Machuca@houstontx.gov
Website: www.fly2houstonspaceport.com

The plan to develop America's tenth licensed commercial spaceport was announced on June 30th, 2015, when the Federal Aviation Administration (FAA) announced it had approved the license for Houston Spaceport. Located at Ellington Airport, Houston Spaceport will offer:

- ➢ Component development and fabrication: much of this will take place in the Houston Aerospace Support Center (HASC—for those who are familiar with NASA, his facility is next door to the Neutral Buoyancy Lab), a 53,000 square foot facility that will feature laboratory and 17,000 square feet of office space. HASC will include:
  - 36,000 square foot high bay
  - Hook crane
  - Two five-ton hook bridge cranes
  - Single hook crane
- ➢ Space vehicle assembly
- ➢ Zero-gravity scientific experiments
- ➢ Microsatellites
- ➢ Astronaut training
- ➢ Space tourism
- ➢ Flexible High Bay layout that will include fabrication and machine shops

Partners

One of the reasons for building a spaceport is to support space tourism, but this business was supposed to have taken off years ago, which is why the FAA's decision to give Houston the green light raised more than a few eyebrows. While Virgin Galactic has sold hundreds of tickets to suborbital space, the date when revenue flights will begin are many years over the horizon, a situation that has already jeopardized the $200 million facility that is Spaceport America that depends on Virgin Galactic as its anchor tenant. The point is that the business of shuttling passengers to space is speculative, which is why it is important for any spaceport to diversify, and this means including a broad range of partners as illustrated by Houston Spaceport's portfolio below:

- ➢ NASA—Johnson Space Center
- ➢ Greater Houston Partnership

- The Sierra Nevada Corporation
- The City of Houston
- Catapult Satellite Applications
- SICSA
- The Boeing Company
- Aerospace Corporation
- Intuitive Machines
- Atec
- Rice Space Institute
- The Texas University Consortium
- Wyle
- CASIS
- UTC Aerospace System

Spaceports in Development: Caribbean Spaceport

www.caribbeanspaceport.com

Caribbean Spaceport
Sphinx Building
Baron G.A. Tindalplein suite #185
1019TW Amsterdam
The Netherlands

Email: info@CaribbeanSpaceport.com
Phone: +31-(0)6 123-66-000 or
Phone: +31 (0)6-506-07-110
Fax: +31 (0)20-776-2775

For those who would like to combine their trip of a lifetime with another destination vacation, there may be no better place than the Netherlands Antilles island of Curaçao which may one day be home to the Caribbean Spaceport (CSP). Originally conceived in 2005 in cooperation with various spaceflight and business professionals, the spaceport is run by Spaceport Partners, which work closely with governmental, academic and business institutions to research and assess the technological, legal and economic feasibility of developing and operating the spaceport. In August 2008, the CSP venture spent 2 weeks on Curaçao to present its plans to government officials, local business people, and the general public. The idea received a warm welcome, prompting CSP founder and director Joost Wouters to extend an invitation to Buzz Aldrin, a group of NASA astronauts, oceanographers and business executives in the SeaSpace group for another presentation in October 2008. This presentation was followed by a February 2009 visit by Sir Richard Branson, who showed great enthusiasm for the CSP concept. Today CSP has concluded its feasibility studies, requirements analyses and business planning and is in the process of discussing investment options and acquiring funding. It is also in contact with various operators and spacecraft developers concerning future operations from its spaceport. When complete, CSP will offer all the facilities necessary for training suborbital passengers, a SpaceExpo, entertainment, bars and restaurants and a shopping mall for friends and family accompanying the space tourist.

Compared to many other spaceport locations, CSP location offers a number of advantages, one of which is using the existing high tech infrastructure of Hato International Airport of Curaçao. Hato's 3.5 km runway is the longest of the Caribbean and is more than long enough to deal with launches for the suborbital spacecraft currently in development. Secondly, unlike many other remotely located spaceports, Curaçao offers an attractive setting with a fully developed tourism infrastructure.

Alternative Revenue Streams #1: Tangential Space-Related Markets

Fig. 9.6 Pedro Duque during training in the Soyuz simulator at Star City. Image courtesy of ESA

Research has indicated that a spaceport that provides only space-specific services will fail [1], which is why it will be important to diversify and create alternate revenue streams. If one examines the airport operations model, it will be noted that 41 % of funding for airport operations is derived from parking revenue, rental car agency revenue, ground transportation services and terminal concessions [1]. While spaceports may not follow the airport model exactly, developing tangential space-related sources of revenue will be a requirement simply for survival. Take Houston Spaceport for example: this spaceport will be in the business of training astronauts *and* building spacecraft among several other alternate sources of space-related activities. The bottom line in making a spaceport profitable is innovation, so what might these tangential space activities be? Well, one source of revenue is to provide visitors and friends and family of those with a ticket to space to vicariously enjoy the experience by participating in some of the training activities listed below.

1. Flight simulators (Fig. 9.6)
2. Centrifuge rides
3. Neutral Buoyancy Lab Indoctrination
4. Parabolic Flights
5. Space science and technology exhibits

Alternative Revenue Streams #2: Space-Related Markets

1. Space weddings
2. Suborbital space diving
3. Commercial filming
4. Advertising: Media sponsorship can be another source of funding as evidenced by the commercial filing performed by RadioShack conducted on the ISS, the $1 million Pizza Hut spent to put its logo on the Proton rocket, and $65 million that a Canadian golf equipment maker paid the Russian Space Agency for a cosmonaut to perform a golf shot outside of the ISS during a spacewalk [2].

Reference

1. FuturIST: Future Infrastructure for Space Transportation. Final Report, International Space University Space Studies Program 2008, pp. 81–94
2. Onuki, M., Lauer, C.J.: SPORTS IN SPACE—another space utilization that drives space commercialization (2007)

Appendix A
Spaceport Functions (Adapted from a document produced by the Vision Spaceport team)

E. Seedhouse, *Spaceports Around the World, A Global Growth Industry*,
SpringerBriefs in Space Development, DOI 10.1007/978-3-319-46846-4

Functions	Description	Sub-functions	Examples of existing facilities and equipment
Flight crew and passenger processing	Functions required at the spaceport to prepare flight crews for their trip	• Medical, emergency, and security services • Flight support equipment integration into vehicle • Training for departure, in-flight, and arrival functions • Ground supplied environmental controls • Development and approval of flight plans • Flight information provisions	• Medical clinic • Planning and stowage of flight crew provisions • Flight suit fitting equipment • Flight simulators • Environmental control and life support systems, shuttle launch pads
	Functions required at the spaceport to prepare passengers for their trip	• Medical, emergency, and security services • Briefing on departure, in-flight, and arrival processes • Ingress into vehicle • Load personal passenger items	This will depend on the type of spaceflight participant: specific facilities will be required for scientists and payload specialists while generic facilities will be required to support those taking a 'joy-ride' into space
Payload processing	Functions required at the spaceport to assemble, integrate, test, vehicle installation, flight, and post flight pre and post flight operations	• Receiving/inspection • Assemble and integrate • Specialized packaging and handling equipment • Verify payload functional • Install payload in flight vehicle • Support payload tests in flight vehicle • Support departure • Post flight removal from vehicle • Post flight test, disassembly, and shipping	• Lifting strong backs • Cranes • Clean rooms and monitoring instruments • Mobile clean room • Specialized electronic test equipment • Nondestructive evaluation and inspection • Propellant and other liquid loading, unloading and cleaning systems
Flight element receipt and acceptance	Functions required at the spaceport to receive the flight element at the spaceport	• Provision of air, land, sea, transport node • Receiving and inspection • Conditioning if required (purging, temperature and humidity control)	• Specialized transportation and handling equipment • Cranes, platforms, doors and handling equipment • Nondestructive inspection and evaluation • Specialized electronic test equipment • Clean rooms and other environmental preservation controls • Fluids support systems (storage, transport, purge, disposal) • Specialized power support systems (60 Hz, DC power) • Specialized communications systems • Controlled storage facilities

Vehicle assembly and integration	Includes the final assembly of the vehicle elements, and the integration of the elements into a total flight ready configuration for launch activities	• Integrate flight element to ground element • Assemble/mate flight elements if required • Perform interface verification • Perform integrated systems test • Perform servicing/close-out if desired • Transfer elements and interface hardware non-flight items to storage location • Transfer flight vehicle to next function	• Specialized handling and transport equipment • Cranes, platforms, doors and other handling equipment • Specialized electronic test equipment • Clean rooms and other environmental preservation controls • Fluids support systems (storage, transport, disposal) • Specialized power support systems (60 Hz, DC power) • System validation check out equipment • Payloads installed in the Orbiter Processing Facility and their integration support equipment • Specialized communications systems
Departure functions	Functions required to service and send the flight vehicle on its mission	• Verify departure facility on-line/functional • Position flight vehicle for departure • Mate with facility/verify functional interfaces • Verify flight systems readiness for departure • Integrate payload and/or personnel module with vehicle/verify functional interfaces • Provide vehicle weather protection • Perform local servicing of commodities and close-out for flight • Perform remote servicing of commodities and close-out for flight • Ingress crew/passengers • Send the vehicle on its mission • Recycle/refurbish departure facility • Service departure facility support systems • Local National coordination with National Airspace System)	• Launch umbilical tower • Sound suppression water system • Mobile service structures • Removable access platforms • Mobile launch platform • Emergency egress system • Weather protection systems • External environmental control • Fluid support systems • Umbilical interfaces from ground to vehicle • Alignment, hold down and release systems • Pre-departure payload integration and validation systems • Runway • Range monitoring systems located within the Spaceport • Command control and monitoring systems • Weather monitoring • Collision avoidance • Verify GPS and TDRSS readiness prior to launch

(continued)

(continued)

Functions	Description	Sub-functions	Examples of existing facilities and equipment
Landing and recovery	Function that includes the arrival or return of a flight element during the course of a space flight. The arrival may be its trip back from space, perhaps a boost or assist stage returning, or a return to Launch Site abort	• Provide landing area(s) • Provide utilities to vehicle at motion stop • Perform safing/check out for return to spaceport • Provide crew/passenger egress capability • Provide down cargo removal capability • Maintain/verify landing facility and systems • Provide ferry facility and fueling capability • Return element to the spaceport • Transfer vehicle to next facility	• Runway • Landing guidance and range safety trajectory detection systems • Command, control, and monitoring systems • Weather monitoring • Post landing purge, vent, drain, safety systems • Payload recovery systems • Passenger transport (mobile stairs, vans) • Ground tow-bars and tugs
Flight element turn-around/ pre-flight functions (turn around of reusables)	Functions required to prepare vehicle for assembly/departure ops Includes all planned and unplanned processing activity to return/ replenish the vehicle to a state acceptable for subsequent departure	• Facility/GSE preps for vehicle turnaround • Gaining access • Payload removal, reconfiguration and installation • Inspection/checkout • Unplanned troubleshooting/repair • Vehicle servicing • Configuring vehicle/systems to support other turnaround functions	• Integrated vehicle health monitoring (shuttle) and analysis • Maintenance inspection, repair and validation systems in Orbiter Processing Facility • OPF fluid support systems (purge, cleaning, verification, emergency venting) • Cranes and other lifting, positioning, and handling equipment • Clean areas for processing and storage • Ground Support Equipment power systems (60 Hz, DC)
Concept-unique logistics functions	Functions required to ship, receive and repair flight line replaceable units (LRU's) and materials	• Propellants (acquisition, storage, distribution, conditioning verification) • Other fluids and gases and unique consumables • LRU replacement hardware (flight and ground systems)	• Fluids, including propellants • Environmental treatment and disposal systems • Storage and staging of consumable repair and replenish part • Calibration, cleaning, and check out of test equipment • Specialized off line labs for inspection and trouble shooting of unplanned events • Engine inspection and overhaul shops • Black box test facilities • Overhaul of mechanical components for hydraulics/ pneumatics • Thermal Protection System support facilities and equipment

Vehicle depot functions	Functions required to perform offline periodic inspections, checkout and upgrade of vehicles	• Vehicle overhaul, inspection/verification, and modification • Modular element overhaul and • Inspection/verification of flight elements • Hot test propulsion hardware	• Specialized facilities and equipment for the test and overhaul of system components and major subsystems
Traffic and flight control functions (led by ARTWG)	Functions required to monitor and control outbound and inbound space traffic, ground element movements within a spaceport	• Commanding (terminate/abort flight and control of flight) • Perform data analysis and decision making • Collect, process, distribute, display, and archive data: • Tracking data • Telemetry data • Surveillance data • Weather data • Provide communications architecture	• Tracking and traffic control of vehicles in launch and landing operational phases (including interface with range), National Airspace System (NAS) • Weather monitoring for decision making • Communications network(s) • Telemetry uplink and downlink • Radar tracking of vehicle trajectories • GPS vehicle tracking
Spaceport support functions	Functions that may be needed to support flight systems processing functions	• Master planning • Quality assurance • Customer relations and public outreach • Shops and labs • Protective services • Records and documentation • Personnel services • Utilities • Roads and grounds • Weather support • Toxic and hazardous waste disposal • Foods services • Communication/information services • Ground transportation services • Environmental compatibility management • Pyrotechnics storage, handling, and disposal • Personal environmental protection equipment • Facility maintenance services and shops • Customs and export control	• Vehicle manifesting and scheduling • Scheduling of maintenance and operations of ground assets • Off site fabrication and repair contractors • Ground systems sustaining engineering • Photographic services • Bonded or secure storage • Cleaning services • Library • Historical archive • Technical documentation center • 60 Hz electric power distribution system • Potable water and fire protection water supply system • Elevators • Vehicle and heavy equipment maintenance shop • Environmental monitoring • Bus services • Cafeteria, snack bars, and vending machines

(continued)

(continued)

Functions	Description	Sub-functions	Examples of existing facilities and equipment
Operations planning and management	Functions that may be needed to support flight/ground systems processing	• Customer relations • Vehicle manifesting and scheduling • Ground systems scheduling and management • Software integration and maintenance • Personnel management • Sustaining operations engineering • Work control • Public affairs • Business management • Advanced planning • Information management and dissemination	• Unique support facilities • Inmarsat communication system (data/voice)
Connecting infrastructure and external community support services	Functions that may be needed to connect to and sustain the overall spaceport enterprise	• Living accommodations • Connecting utility infrastructure • Transportation support • Educational support • Community police/fire protection • Community/private resources infrastructure and services • Consumer retail support • Community medical support/hospitals, etc • Financial institutions • Economic development • Area environmental support • Emergency preparedness and disaster planning	• Personnel and visitor comfort facilities (homes, hotels) • Community facilities (stores, schools, churches, library, entertainment, restaurants, local government, utilities, roads) • Ecological environment • Power substations • Visitor's center • Spaceport gates (controlled access) • Planning for hurricane, earthquake, terrorist attack, etc

Appendix B
Outer Space Treaty

- ➢ This provides a framework for international space law that includes the following principles:
  - λ Exploration and use of outer space shall be carried out for the benefit and in the interests of all countries and shall be the province of all mankind
  - λ Outer space shall be free for exploration and use by all States
  - λ Outer space is not subject to national appropriation by claim of sovereignty, by means of use or occupation, or by any other means
  - λ States shall not place nuclear weapons or other weapons of mass destruction in orbit or on celestial bodies or station them in outer space in any other manner
  - λ The Moon and other celestial bodies shall be used exclusively for peaceful purposes
  - λ Astronauts shall be regarded as the envoys of mankind
  - λ States shall be responsible for national space activities whether carried out by governmental or non-governmental entities
  - λ States shall be liable for damage caused by their space objects
  - λ States shall avoid harmful contamination of space and celestial bodies.

- ➢ The Outer Space Treaty Basic framework on international space law, includes the following principles:

  - Liability Convention λ Elaborates on Article 7 of the Outer Space Treaty which establishes liability of launching state λ Launching State shall be: λ absolutely liable to pay compensation for damage caused by its space objects on the surface of the Earth or to aircraft, and λ liable for damage due to its faults in space. λ The Convention also provides for procedures for the settlement of claims for damages.
  - Registration Convention λ Launching State should furnish to the U.N. the following information concerning each space object: λ Name of launching State; λ An appropriate designator of the space object or its registration number; λ Date and territory or location of launch; λ Basic orbital parameters,

E. Seedhouse, *Spaceports Around the World, A Global Growth Industry*,
SpringerBriefs in Space Development, DOI 10.1007/978-3-319-46846-4

including: λ Nodal period (the time between two successive northbound crossings of the equator—usually in minutes); λ Inclination (inclination of the orbit—polar orbit is 90° and equatorial orbit is 0°); λ Apogee; λ Perigee λ General function of the space object.

- Rescue and Return Agreement λ Elaborates articles 5 and 8 of the Outer Space Treaty λ States shall take all possible steps to rescue and assist astronauts in distress and promptly return them to the launching State, and λ States shall, upon request, provide assistance to launching States in recovering space objects that return to Earth outside the territory of the Launching State.
- Export Controls and International Cooperation λ International Traffic in Arms Regulations λ USML Category XV λ Technical Assistance Agreements λ Technology Transfer Control Plans λ Relevant Licensing Offices λ Directorate of Defense Trade Controls λ Defense Technology and Security Administration λ Congressional Notification.

Appendix C
§ 420.15 Information Requirements

General

(1) *Launch site operator*. An applicant shall identify the name and address of the applicant, and the name, address, and telephone number of any person to whom inquiries and correspondence should be directed.
(2) *Launch site*. An applicant shall provide the name and location of the proposed launch site and include the following information:

   I. A list of downrange equipment;
   II. A description of the layout of the launch site, including launch points;
   III. The types of launch vehicles to be supported at each launch point;
   IV. The range of launch azimuths planned from each launch point; and
   V. The scheduled operational date.

(3) *Foreign ownership*. Identify foreign ownership of the applicant, as follows:

   I. For a sole proprietorship or partnership, all foreign owners or partners;
   II. For a corporation, any foreign ownership interest of 10 % or more; and
   III. For a joint venture, association, or other entity, any foreign entities participating in the entity.

Environmental

An applicant shall provide the FAA with information for the FAA to analyze the environmental impacts associated with the operation of the proposed launch site. The information provided by an applicant must be sufficient to enable the FAA to

E. Seedhouse, *Spaceports Around the World, A Global Growth Industry*, SpringerBriefs in Space Development, DOI 10.1007/978-3-319-46846-4 comply with the requirements of the National Environment Policy Act, 42 U.S.C. 4321 *et seq.* (NEPA), the Council on Environmental Quality Regulations for Implementing the Procedural Provisions of NEPA, 40 CFR parts 1500–1508, and the FAA's Procedures for Considering Environmental Impacts, FAA Order 1050.1D. An applicant shall submit environmental information concerning a proposed launch site not covered by existing environmental documentation, and other factors as determined by the FAA.

Launch Site Location

(1) Except as provided by paragraph (c)(2) of this section, an applicant shall provide the information necessary to demonstrate compliance with §§ 420.19–420.29.
(2) An applicant who is proposing to locate a launch site at an existing launch point at a federal launch range is not required to comply with paragraph (c)(1) of this section if a launch vehicle of the same type and class as proposed for the launch point has been safely launched from the launch point.

Explosive Site Plan

(1) Except as provided by paragraph (d)(2) of this section, an applicant shall submit an explosive site plan that complies with §§ 420.63, 420.65, 420.67, and 420.69.
(2) If an applicant plans to operate a launch site located on a federal launch range, and if the applicant is required by the federal launch range to comply with the federal launch range's explosive safety requirements, the applicant shall submit the explosive site plan submitted to the federal launch range.

Launch Site Location

An applicant shall provide the information necessary to demonstrate compliance with the requirements of §§ 420.53, 420.55, 420.57, 420.59, 420.61, and 420.71.

Appendix D
Current U.S. Liability Risk-Sharing Regime Under 49 U.S.C. Subtitle IX, Chapter 701

The U.S liability risk-sharing regime for commercial space transportation is comprised of three tiers:

Tier I: Maximum Probably Loss (MPL)-Based Financial Responsibility Requirements

- Launch or re-entry license obtains insurance to cover claims of third parties, including Government personnel, for injury, loss or damage, against launch or re-entry participants. Participants include the licensee, its customer, and the U.S. Government and its agencies, and the contractors and subcontractors of each of them.
- Launch or re-entry licensee obtains insurance covering damage to U.S. Government range property.
- The Federal Aviation Administration (FAA) sets insurance requirements based upon the FAA's determination of the MPL that would result from licensed launch or re-entry activities, within statutory ceilings, not to exceed the lesser of:
  - $500 million for third-party liability, or the maximum allowable on the world market at reasonable cost.
  - $100 million for U.S. Government range property, or the maximum allowable on the world market at reasonable cost.
- Participants enter into no fault, no subrogation reciprocal or cross-waivers of claims under which each participant accepts its own risk of property damage or loss and agrees to be responsible for injury, damage or loss suffered by its employees, except that claims of Government personnel are covered claims under the licensee's liability insurance coverage.

E. Seedhouse, *Spaceports Around the World, A Global Growth Industry*,
SpringerBriefs in Space Development, DOI 10.1007/978-3-319-46846-4

Tier II: Catastrophic Loss Protection (Government Payment of Excess Claims, Known as 'Indemnification')

- Subject to appropriations, the U.S. Government may pay successful third-party liability claims in excess of required MPL-based insurance, up to $1.5 billion (as adjusted for post-1988 inflation) above the amount of MPL-based insurance.
- U.S. Government waives claims for property damage above required property insurance.

Tier III: Above MPL-Based Insurance Plus Indemnification

- By regulation, financial responsibility remains with the licensee, or legally liable party.

Index

E. Seedhouse, *Spaceports Around the World, A Global Growth Industry*, SpringerBriefs in Space Development, DOI 10.1007/978-3-319-46846-4